RESOURCE RECOVERY ECONOMICS

POLLUTION ENGINEERING AND TECHNOLOGY

A Series of Reference Books and Textbooks

EDITOR

PAUL N. CHEREMISINOFF

Associate Professor of Environmental Engineering
New Jersey Institute of Technology
Newark, New Jersey

Additional Volumes in Preparation

RESOURCE RECOVERY ECONOMICS

Methods for Feasibility Analysis

STUART H. RUSSELL

Hennington, Durham, and Richardson
Omaha, Nebraska

MARCEL DEKKER, INC.   New York and Basel

Library of Congress Cataloging in Publication Data

Russell, Stuart H.,
    Resource recovery economics.

    (Pollution engineering and technology ; 22)
    Includes index.
    1. Salvage (Waste, etc.)--Cost effectiveness.
2. Recycling (Waste, etc.)--Cost effectiveness.
3. Conservation of natural resources--Cost effectiveness.
4. Energy conservation--Cost effectiveness. I. Title.
II. Series.
HD9975.A2R83 1982   363.7'28   82-18232
ISBN 0-8247-1726-0

MARCEL DEKKER, INC.
270 Madison Avenue, New York, New York 10016

Current printing (last digit):
10 9 8 7 6 5 4 3 2 1

PRINTED IN THE UNITED STATES OF AMERICA

PREFACE

Around the time of the 1973-1974 oil embargo, those of us in the solid waste management business began to hear and read with increasing frequency that the halcyon days of the "throw-away" mentality were gone forever to be replaced by more parsimonious use of resources and by the recovery of materials and energy from wastes. Although resource recovery was touted as the answer to both waste disposal and energy problems, it never seemed to garner the wide-spread acceptance which seemed so certain. Hundreds (perhaps thousands) of feasibility studies were performed, but only a few such studies resulted in operating facilities. In some communities, studies were done over, and over again in a seemingly interminable series of analyses of different options under different assumptions. Many times, facilities which were implemented did not perform as predicted, and most did not meet the economic projections contained in the feasibility studies. There are many political, economic, and technical reasons for the slowness with which the resource recovery industry has developed. One major problem has been the improper preparation of initial feasibility analyses which have either promoted bad projects or stopped good ones.

Now, almost ten years later, it appears that resource recovery has finally come of age. (In some areas of the country there is actually competition among proposed resource recovery projects for the solid waste generated in a community.) The continuous rise in energy costs, strident citizen opposition to the siting of new landfills, and federal incentives for the development of alternative energy sources have pushed resource recovery in many communities over the economic feasibility "hump." Resource recovery will not be feasible for every community, but it is now more important than ever to investigate the

iii

feasibility of resource recovery as a component in a community's solid waste management system. The purpose of this book is to give the reader the benefit of the experience of many people in the solid waste management field who have performed feasibility studies for resource recovery. A method developed over years of performing such studies in almost every type of community is presented to guide the reader through the steps necessary to perform a proper feasibility study.

The objective of the feasibility study methodology presented in this book is to avoid false starts and multiple studies by examining all available options on an equivalent basis. The study method gives the municipal official a structure, within which conflicting information from system vendors, potential private developers, and citizens can be reconciled in a comprehensive manner. The methodology is a "systems approach." That is, all elements of the solid waste management system (waste transport, processing, combustion, final disposal) are examined rather than individual facilities (landfills, transfer stations, resource recovery plants, etc.). The method calls for selecting a set of "system alternatives" and making a comprehensive economic comparison of them as the basis for a decision. Previously unpublished data from actual feasibility studies are presented to aid the reader in making comparisons. A hypothetical community, River City, is used as a numerical example to illustrate the required analyses in each of the study steps.

This book is dedicated to those municipal officials who wish to compare, without bias, the economics of resource recovery versus landfilling and other Conventional disposal methods. Conventional disposal methods may well prove more economical in many cases. It is hoped, however, that with a clear, logical, and comprehensive economic analysis, the quality of the resource recovery systems which prove feasible in the study will be higher in terms of technical and economic viability when built, than those based on an incomplete and unclear analysis. In the long run, quality is better than quantity for the resource recovery industry.

Stuart H. Russell

ACKNOWLEDGMENTS

This book is the result not only of the author's efforts, but of the talents and efforts of a number of the author's colleagues in the Solid Waste/Resource Recovery Division of Henningson, Durham, & Richardson (HDR). The entire group of engineers and scientists who work in this division of HDR contributed to the book over its gestation period. The book has benefited greatly from each individual's experience and insight.

Those who deserve special recognition include the division director, Harvey Funk, who has supported the effort of writing this book over the last two years, Frank Borchardt, Mary Wees, Richard Bell, Russell Menke, Bill Hazell, and Dave Maystrick. Special thanks also to Wendy Thran and Sue Coffee for help in preparing the manuscript.

CONTENTS

CHAPTER 1

INTRODUCTION

I. PURPOSE

Every municipal official concerned with solid waste management is confronted with questions from citizens, industry, and public officials about solid waste resource recovery every day:

"Why don't we burn all of our garbage in a boiler and make electricity?"

"I read somewhere that some cities are making fuel out of their solid waste. Should we be doing that in our community?"

"Will resource recovery eliminate the need for our landfill?"

"Why can't we burn all of our waste to make steam for our local industry?"

"We should be composting all of our waste. Why aren't we doing this in our community?"

"Why don't you buy my resource recovery equipment? It will make money for your community."

"I was told that resource recovery has never been tried before. Shouldn't we just forget about it for our community?"

1

"Isn't it true that resource recovery is too expensive and has
not worked in the existing plants around the country?"

These questions can be ignored or answered in a haphazard manner, but
definitive answers will eventually be required. The demand for answers
to these questions has been the result of a dramatic increase in
interest in solid waste management and in the concept of recovering
energy and materials from solid waste before disposal. This interest
has been stimulated by an increased awareness of the environmental and
economic problems associated with past disposal methods, and by the
widespread shortages and increased prices of energy and materials.

In the past, town solid waste managers, city public works directors,
or county solid waste officials have had only to deal with the most
economical way to landfill solid wastes. As a result of this recent
increase in interest in solid waste management, these same public
officials are now being asked to make complex analyses of the various
alternative disposal methods to determine which is best. Lack of
information, the presence of inaccurate information and misconception,
and most importantly, the lack of a sound methodology for performing
these complex analyses has made the question of resource recovery
foggy and confused.

The many considerations which go into selecting a solid waste
disposal or resource recovery system for a given community can be put
into three categories:

Technical/Economic   -   What equipment has been proved?
                       -   Is it reliable?
                       -   Are there buyers for its products?
                       -   Which equipment would best serve the
                           buyers?
                       -   How much does it cost to build?
                       -   How much does it cost to operate?
                       -   How does its cost compare with other
                           alternatives?
                       -   How can it be financed?

Environmental   - How will it affect ambient water, air,
                         noise, etc.?
                       - Will it improve or degrade the
                         environment?
                       - Will it create aesthetic problems in its
                         vicinity?
                       - Is it worth paying more money for
                         environmental improvement?

Political/Social   - is there political support for the
                         concept?
                       - will citizen groups allow siting?
                       - is the money to build it best spent
                         elsewhere?

All of these questions must be resolved when making a sound decision.
Each community has a different set of conditions, but in every case an
economic analysis is the essential first step in the decision-making
process.   The equipment and its outputs must be defined before
environmental analyses can be made.   The plant or landfill site and
cost must be defined before political support can be assessed.   Indeed,
many times the "dollars" issue is the sole determinant of a solid waste
disposal or resource recovery system's viability.

It is the purpose of this book to focus on this economic analysis
step of the decision-making process.   This initial economic analysis step
is many times called a "Feasibility Study."   The thrust of the following
chapters is not to tell the reader which resource recovery systems are
the least expensive.   This task is impossible because a sound economic
analysis must include a variety of factors about a given community
which can only be assessed by those in that community.   The main
purpose, rather, is to detail an economic analysis methodology for the
public official or solid waste task force to follow.   This methodology is
a step-by-step procedure which can be used to sort out and reconcile
conflicting information so that valid comparisons of the alternatives can
be made.

With this methodology, many of those daily questions about solid waste disposal and resource recovery can be answered. The methodology can be used to support the continued operation of a local landfill, or if local conditions are favorable, to support the push for resource recovery.

The methodology covered in this book will result in the recommendation of a solid waste management system alternative (which may or may not include resource recovery) based on many initial assumptions. It is generally preferable for a community to form a resource recovery "task force" or "committee" comprising key decision-makers in the community to assist in the analysis and in making required assumptions. Many times members will include the person responsible for solid waste collection and disposal in the community, other public works personnel, a director of financial matters for local government, a representative of the applicable planning department and legal department. These members can represent a number of interested public entities (towns, villages, cities, counties, states) or a single entity. Private citizens may also be made part of the task force (e.g., members of environmental groups, private waste collection companies, representatives of major industries).

It is important for this committee or task force to take action on two matters before beginning the analysis. First, a project coordinator should be appointed. One person should be responsible for coordinating the results of all analyses and presenting the conclusions to the task force and others. This person must be committed to following through with project over a period of one year or more. The committee should also define its goals as soon as possible. Questions such as the following should be answered:

Is cost the only consideration when choosing a solid waste system, or will environmental enhancement, energy conservation, or other considerations be factors?

What political jurisdictions are to be involved in the feasibility study?

Is the primary purpose for considering resource recovery the disposal of solid waste, or might there be other purposes as well, such as economic development (attracting new industry), or becoming a primary energy producer?

How much control does local government wish to have over the operation of the resource recovery system?

Definitive answers to these and other questions early in the study will avoid time-consuming and costly misdirection of the economic analysis.

II.   DEFINITIONS

Any area of endeavor from biochemistry to skiing has its own set of terminology. Solid waste management and resource recovery has a terminology which has changed as rapidly as the technology in recent years. For this reason, many terms are ill-defined and confusing. The terms relative to the areas covered by this book will be defined in each individual chapter; however the definition of two general terms at this point in the book will be helpful in defining the scope of coverage. These two terms are "Solid Waste" and "Resource Recovery."

A.   Solid Waste

Solid waste goes by many names: refuse, garbage, MSW, rubbish, etc. A number of literature sources (1,2) give in-depth discussions of the nature of solid wastes. These references should be consulted for more detailed definitions. It is generally accepted that the term "solid waste" is a broad term encompassing all solid and semi-solid discarded materials. Although existing references use various categorizing schemes, solid waste can generally be divided into the following categories:

<u>Residential</u> - These wastes are often called "domestic" or "household" wastes. Wastes in this category consist of those which are normally collected from single and multiple family residences in a municipal collection program.

<u>Commercial</u> - These wastes come from a variety of commercial businesses engaged in wholesale and retail trade; finance, insurance, and real estate; health care and other services; and government. Infectious wastes from health care establishments or bulky wastes such as discarded equipment are not included. These are included in the "special" wastes category.

<u>Industrial</u> - This group include wastes from businesses engaged in manufacturing, transportation, communication, and public utilities. This is a highly-diversified group of wastes which can include paper, wood, and metal from packaging and office activity; food wastes from cafeterias; "special" industrial wastes such as fly ash, bottom ash, air pollution control solids, foundry sand, blast furnace slag, and others; and "Hazardous Wastes" as defined by 40 CFR Part 261 which include certain sludges, scrap chemicals, spent solvents, spent electroplating baths, and many other ignitable, corrosive, reactive, or toxic wastes.

<u>Special</u> - This is another highly-diversified group of wastes including construction/demolition debris, street sweepings, bulky wastes, infectious wastes, recreational wastes, junked automobiles, and tires.

<u>Agricultural</u> - These wastes consist primarily of live-stock and poultry manure, crop residue, and dead animals.

When considering resource recovery, a community should generally be concerned with the first two categories plus the packaging, office and cafeteria part of the Industrial waste category. Although these combined wastes can be as little as 30 percent to 40 percent of total solid waste generation, these are the only categories which can practically be used in a municipal resource recovery system. The term "processable" will be used in this book to describe these wastes. Most mechanical resource recovery equipment has been designed to process only these waste fractions which consist of about 78 percent paper and other organics, about 10 percent glass, about 10 percent metals, and other miscellaneous inorganics (3). Special, Agricultural and the non-processable Industrial wastes have widely varying compositions and/or a hazardous nature which cause handling and processing problems. In addition, since the processable waste fraction is normally collected in a municipal waste collection program by either public workers or contractors, the municipality usually has a certain measure of control over where the waste is deposited and data on how much is collected. Much of the non-processable Industrial, Special, and Agricultural wastes are handled and disposed by individual generators so these wastes never enter the municipal waste stream.

For these reasons, when the term "solid waste" is used in this book, the reference is to the processable waste fractions of the total waste stream unless otherwise noted.

B. Resource Recovery

To many, the term "resource recovery" means a central, mechanical processing facility which separates materials and produces energy from solid waste. The term is ambiguous, however, because it is unclear whether recycling, baling, composting, landfill gas extraction, or others also are included in the term. Using its broadest interpretation, resource recovery means," any materials separation and/or chemical conversion of solid waste for the purpose of recovering materials or energy products." This definition includes manual separation by individual residents before collection (source separation), landfill gas extraction after disposal, as well as the various methods of mechanical processing and energy conversion between collection and disposal.

Although these pre-collection and post-disposal activities can be an important part of a complete solid waste management program, this book will be concerned mainly with resource recovery which takes place between collection and disposal. Included are such technologies as mechanical separation and size reduction, waterwall incineration, combustion of prepared fuel, pyrolysis, and others. Rather than focusing only on a separation and/or combustion plant, however, this book will concentrate on "resource recovery systems" which may include combinations of separation plants, combustion plants, transfer stations, a transport network, and landfills. Resource recovery cannot be analyzed in a vacuum. The various technologies for separation and combustion must be combined with the more conventional methods of solid waste management in order to form a system which will handle all of the processable solid waste from the end of the collection route through final disposal. This "total sytem" approach will be discussed in more detail in the following chapters.

III. ECONOMIC ANALYSIS METHODOLOGY

The economic analysis methodology discussed in this book is illustrated in simplified form in Figure 1.1. This logic block diagram shows how the various activities which compose the economic analysis interrelate. Note that input from environmental and political considerations has been shown to be part of the selection of the best alternative. The following sections are short discussions of the parts of this economic analysis which are covered by each chapter of the book.

A. Basic Data

The methodology starts with the gathering of the basic data necessary to initiate a resource recovery study. Data such as population, employment, transportation time and distance, existing landfill disposal records, landfill and transfer station sites are used with a variety of calculation methods to arrive at 1) present and future waste quantities, 2) present and future waste composition, and 3) present and future

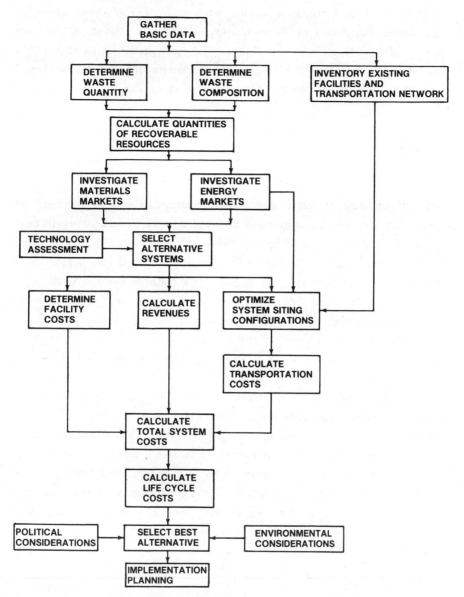

Figure 1.1. Economic analysis methodology, block diagram.

facility sites and transportation networks. These three areas of information illustrated by three blocks in Figure 1.1 must be derived before the analysis can continue. The data gathered in this initial activity will determine the results of the entire analysis. Chapter 2 gives the methods and sources for gathering these data.

B. Markets

Once the amount of waste and its composition have been defined, the amounts of recoverable materials and energy can be estimated through an initial screening of resource recovery technologies so that discussions with potential markets (buyers of recovered resources) can begin. This step is absolutely necessary for the analysis because there is little logic in recovering materials and energy if there are no buyers, or spending time investigating buyers for materials or energy which will not be produced.

The identification of local energy markets is the most important market analysis task because energy sales account for the majority of a system's revenues. Materials recovery alone has not proved financially viable. Potential buyers of energy may include local industries, institutions, utilities, or district heating systems. Each potential market must be investigated to determine the amount and type of energy product (i.e., solid fuel, steam, electric power, etc.) which can be utilized, energy product specifications, seasonal demand variations, type of existing energy conversion equipment, and the potential market value of recovered energy. Chapter 3 shows how to conduct a thorough local market survey.

C. Alternatives

The market surveys, especially the energy market investigations, determine the types of energy and material products which are marketable. This part of the economic analysis methodology involves a more detailed assessment of existing resource recovery technology to determine the technologies available for producing the specifications required by the markets. An assessment of technical viability, operating experience, and equipment reliability must be made in this analysis. The object of this task is to select a set of system alternatives for economic comparison (see Figure 1.1). This selection must be carried out carefully because the alternatives selected must have comparable solid waste handling capacities and operating lifetimes. Each system alternative must be a combination of transport, resource recovery, and/or landfill systems which handle either the entire waste stream, or equivalent portions of the waste stream (processable and non-processable). If one alternative out of the selected set handles a different amount of solid waste than other alternatives, accurate economic comparison is not possible.

It is also imperative that a "landfilling-only" alternative be included in this set of system alternatives for a comparison baseline. A landfilling system is not comparable unless it has an operating life equal to that of the resource recovery systems. It may therefore be necessary to project existing landfill capacity and select future landfill sites for inclusion in the analysis. Techniques and information used in making the selection of system alternatives are discussed in Chapter 4.

D. Costs

This section of the analysis is concerned with making accurate estimates of the capital and annual costs of the facilities needed for each of the selected system alternatives. Costs are required for transfer stations and landfills as well as for resource recovery facilities. It is important in this step to include all of the costs which will actually be incurred. For example, the costs for engineering, support equipment, and financing have been neglected in many cost estimates. Chapter 5 describes the cost components which should be included along with typical cost estimating factors and sources of information.

E.   Systems Analysis and Life Cycle Costs

The object of this section of the economic analysis is to bring together all of the cost components which make up "net system cost." This requires the calculation of revenues from the market information and the calculation of transportation costs by optimizing resource recovery facility, transfer station, and landfill locations using the siting and transportation network data developed in the Basic Data section. Combined with the facility costs previously calculated, the net cost from the end of the collection route through final disposal can be calculated for each alternative. The effects of lower transportation costs due to centralized facilities can be shown using this approach. Computer models which perform this type of analysis will be described in Chapter 6, but the emphasis will be on showing a simplified, hand-calculation approach which can be used where computer modeling may not be practical.

Another essential part of this economic analysis methodology is to perform a life cycle cost analysis for each alternative. Net system costs are not static, they change over time. The rising costs of labor, materials and, most importantly, energy will assure these changes. The life cycle analysis is a projection of the total system costs of each alternative over the system's operating life. The life cycle cost analysis can show if there is the potential for an alternative system which has a presently high cost to become a lower cost alternative due to revenue increases related to rising energy and materials values. It is important in making the final selection to know if this potential exists. Chapter 6 will detail the method and typical projection factors which can be used in such an analysis.

F.   Implementation Planning

This is the final step in the study of a resource recovery system. After an alternative system has been selected using the economic, political, and environmental analyses, a plan for implementing the recommended system must be presented to the appropriate decision-makers. This plan must show the steps necessary to move from the end of the feasibility study to the construction and operation of the system along with the time and funding required. Chapter 7 presents a discussion of some of the justifications and impediments to implementation along with a generalized planning framework to be used in developing an implementation plan.

IV. HYPOTHETICAL STUDY AREA: RIVER CITY, U.S.A.

This economic analysis methodology is more easily understood if it is applied. For this reason, and to provide continuity between chapters, a hypothetical study area called River City has been postulated. The River City metropolitan area will be used to illustrate in each chapter how each part of the methodology can be applied to an example situation.

River City is a medium-sized metropolitan area of about 588,000 residents with a diversified economic structure. Principal economic activities include agricultural trade, service and processing; light manufacturing; transportation; insurance; and utilities. Growth and residential expansion has been concentrated mainly on the west side of the river. Population growth has been steady, but not overly rapid--about 2 percent per year.

The climate of River City is marked by seasonal variations in temperature and precipitation. Mid-summer daytime temperatures average 85 to 87 degrees while daytime winter temperatures average 31 to 32 degrees, with nighttime lows of 15 to 19 degrees.

Figure 1.2 shows the existing solid waste facilities in River City. All of the residential solid waste is presently collected by a hauling contractor paid by the city. One transfer station is operated by the city which transfers waste collected mainly in the central business district to the central landfill. The remainder of the generated waste (commercial, industrial and special) is handled under individual contracts with private haulers or by the city (street sweepings, etc.).

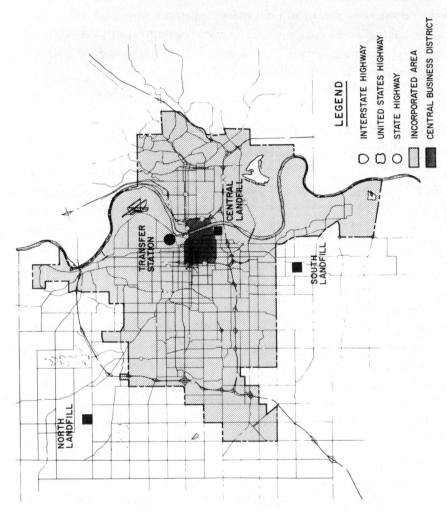

Figure 1.2. River City, existing solid waste facilities.

There are three major landfills in the River City area with characteristics as given in Table 1.1. Note that the only landfill with a substantial remaining life (ten years) is the county-owned North Landfill. At increased fill rates resulting from the closure of the other landfills, however, its remaining life may be shorter than predicted. Recent attempts to locate a new city landfill have met with heavy opposition from area residents. The city is therefore faced with the difficult choice between locating a new landfill in the city or transporting waste to an outlying part of the county where a new county landfill may be approved.

Because of these looming landfill problems, the River City Department of Public Works has become interested in investigating the possibility of resource recovery and how the costs of such a system would compare to the landfilling alternatives.

The following chapters will use the River City example to illustrate each step of the economic analysis process which will ultimately result in the choice of the best solid waste management system.

TABLE 1.1

Existing Landfills in River City

Landfill	Ownership and Operation	Remaining Capacity (years)
North Landfill	County	10
Central Landfill	City	3
South Landfill	Private	7

REFERENCES

1. D.J. Hagerty, J.L. Pavoni, and J.E. Heer, Jr., Solid Waste Management, Van Nostrand Reinholt, New York, 1973, pp. 3-22.

2. W.R. Niessen in Handbook of Solid Waste Management D.J. Wilson, ed., Van Nostrand Reinholt, New York, 1977, pp. 10-61.

3. U.S. Environmental Protection Agency, OSW, Fourth Report to Congress--Resource Recovery and Waste Reduction, SW-600, Washington, D.C., 1977, p. 14.

CHAPTER 2

BASIC DATA

I.   GENERAL

Gathering a complete set of reliable basic data is the critical first step in the economic analysis of resource recovery. The accuracy and viability of all calculations performed in the other sections of the analysis depend on the data gathered in this task. For example, waste quantity information is the basis for facility sizing (and hence capital and operating costs), transportation costs, and the calculation of revenues.

A solid waste management system has many elements. Figure 2.1 shows that these elements may include collection, primary haul, transfer, processing/recovery, secondary haul, and final disposal (1). Except for the collection element, all elements of this system may be changed by the implementation of a resource recovery facility as an alternative to landfilling. For example, a resource recovery facility location separate from the landfill location will change waste transportation distances and costs. Because a certain portion of the waste stream is diverted from land disposal by the resource recovery facility, final disposal quantities and costs are reduced. The introduction of a resource recovery facility into a system in which none previously existed will obviously create processing/recovery costs and revenues.

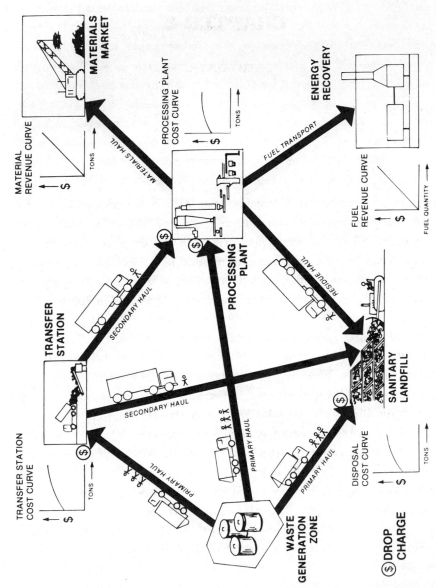

Figure 2.1. Elements of Solid Waste Management

The basic data required to analyze the system cost implications of these changes must be gathered in this task. Data relating to existing and projected waste quantity and composition must be gathered. The areal distribution of waste generation must be known. Potential sites for new landfills, transfer stations, and resource recovery facilities must also be identified. The unit costs of hauling wastes are also critical.

Before beginning the data gathering task, the investigator must establish the level of accuracy required. Different types of analyses require different levels of accuracy. For example, if the results of the economic analysis are to be used as the basis for issuing $100 million of solid waste disposal bonds, the data used should be as accurate as possible. If, however, the results are to be used to establish only initial feasibility in anticipation of further study, the data gathering task can be simplified.

The required level of accuracy significantly influences the cost of data collection. An extensive, year-long landfill sampling and analysis program to establish solid waste quantity and composition certainly will cost more than utilizing existing scale data and composition information. The results of the economic analysis are only as accurate as the basic data, but the investigator must balance the costs of data collection against the intended use of the results. The type of analysis detailed in this book is an initial feasibility determination. It is intended to provide the basis for selecting a certain system for further analysis before implementation. As such, a level of data accuracy appropriate for such an analysis is assumed.

The following sections will discuss the methods which can be used to develop a complete set of basic data for a feasibility analysis. The final section of this chapter will apply these methods to the hypothetical River City case for numerical illustration.

II. DATA REQUIREMENTS AND SOURCES

Before beginning the data gathering task, the study area must be defined. This may seem a trivial task, but the study area boundaries should be carefully selected. In the case of a city study, for example, it may be advisable to include certain surrounding communities or unincorporated areas which use the same landfills. In certain large cities, it may be necessary to divide the city into sections for separate analysis in order to reduce the required facility sizes and hauling distances. In certain counties or states, it may prove to be more efficient to limit the study area boundaries to the regions with the highest population densities. In all situations it is important to define explicitly the study area boundaries in advance, so confusion in the data gathering step can be avoided.

A. Waste Generation

Perhaps the single most important set of basic data for a resource recovery feasibility analysis is the existing and projected quantity and composition of the processable solid waste which must be disposed in the study area. It is also perhaps the most difficult data to accurately define in many cases.

1. Existing System

An examination of the existing collection/haul/disposal system is the necessary first step in defining waste generation. The investigator should identify the major private and municipal collectors of residential, commercial and industrial waste in the study area. The existing landfill locations in which waste collected in the study area is disposed should be identified and plotted on a map. If possible, other existing data about each landfill should be gathered, including 1) present incoming waste quantity rates and a breakdown of the areas from which the wastes come, 2) expected remaining operating life and/or acres remaining in the fill plan, 3) anticipated expansion plans, 4) existing environmental problems or recent enforcement actions against the operators, 5) the types of waste accepted, and 6) existing drop charge rate schedules and plans for future increases in rates. Locations and sizes of any other existing facilities, such as transfer stations and incinerators, should also be identified.

This information about the existing system will begin to form the basis of the landfill system baseline case for comparison with the resource recovery alternatives. Most of these facts can be gathered by communicating directly with the landfill operators or the local solid waste department. It may also be necessary to contact the local haulers and the county or state regulatory agency responsible for landfill permits and enforcement.

2. Existing Quantities

In a resource recovery analysis it is necessary to work with units of weight (tons or pounds) rather than volume (cubic yards). Because of the widely varying density of solid waste, a unit of weight gives a much more precise measure of quantity than a unit of volume.

The highest quality data for use in estimating the existing quantity (weight) of waste disposed is landfill or transfer station scale data. Scale data in tons per month should be tabulated for all landfills and transfer stations utilized by the study area. If a certain landfill also accepts wastes from areas outside the study area, the data must be adjusted accordingly. Since scale records are kept for billing purposes, a landfill operator usually will have tonnages recorded for each hauler. By examining the hauler breakdown and by determining the locations of the major collection routes for each hauler, an acceptable adjustment for the study area can be made.

It is also necessary to determine which portion of the total tons received at each landfill or transfer station is processable (residential, commercial, and certain industrial wastes). This data adjustment can also be made by examining the hauler breakdown and by talking to the landfill or transfer station operator. The records will often include an indication of which loads were bulky wastes or construction/demolition debris, because different drop charge rates often apply for these loads.

Also, the operator will probably know which haulers typically bring in non-processables such as slag, ash, etc. Determining the processable waste portion is not an exact science. By making sound assumptions and by communicating with a cooperative operator, however, an acceptable adjustment can be made. The proportion of the total waste which is non-processable is also an important piece of data. Although these wastes cannot be recovered, their quantity will affect future landfill sizing because they must be disposed (except for hazardous wastes which must be disposed in specially-permitted sites). Assumptions and calculations should be clearly documented so that future adjustments can be made if necessary.

If scales are not present at certain landfills, records of numbers of loads, and/or volume are often kept. Although not as accurate for determining the tonnage disposed, these records can be used with appropriate density factors (pounds per cubic yard in the collection vehicle) for calculating tonnage. Density factors should be carefully chosen. Packer manufacturer specifications for achievable densities are usually too high for this calculation. These specifications are generally maximums for the particular packer equipment and do not accurately represent the lower fleet average compaction which occurs in normal operation. Also, the landfill drop charge rate structure frequently will affect the average compaction factor. For example, if landfill billing is on the basis of volume (i.e., dollars per cubic yard or dollars per load), average compaction factors are likely to be higher because of the incentive for the collector to include as much material as possible in each load. If, on the other hand, there is no drop charge (as in the case of some municipal landfills), density factors will likely be lower.

Table 2.1 gives average density factors calculated for different truck types in different areas of the country (2-5). These factors were calculated from truck weighing programs and scale data at landfills. It is advisable to conduct a local truck weighing program to determine accurate density factors for use with local landfill volume records. Portable axle scales can be used with an acceptable degree of accuracy. In such a weighing program, it is important to calculate

TABLE 2.1

Collection Vehicle Compaction Factors for
Mixed Trash and Refuse

Location	Sample Year	Number of Samples	Truck Type	Hauler	Avg. Compaction Factor(lb./CY)	Reference
Omaha, NE	1969	432	RL Packer[a]	Private	464	(2)
		119	Open Body[b]	Private	232	
Dallas, TX	1975	63	RL Packer	Private	450	(3)
		468	FL Packer[c]	City	375	
Fairfax County, VA	1979	41	RL Packer (18 CY)	Private	625	(4)
		67	RL Packer (20 CY)	Private	548	
		52	RL Packer (23 CY)	Private	414	
		41	RL Packer (25 CY)	Private	466	
		480	RL Packer (25 CY)	City	100	
		156	RL Packer (27 CY)	Private	366	
Phoenix, AZ	1976-79[d]	72,807	Packer w/Rapid Rail (24 CY)	City	472	(5)
		68,236	RL Packer (25 CY)	City	377	
		52,373	Shu-Pak (33 CY)	City	309	

a Rear-Load Packer.
b Pick-up Trucks and larger, non-compacted vehicles.
c Front-Load Packer.
d Averages of Landfill scale records over four years of operation.

density factors for different truck types (e.g., pick-up, rear loader, front loader, open body). A sample size of ten percent of the average weekly number of loads for each truck type should be sufficient for data accuracy. The average density factors thus calculated should then be applied to the total volumes disposed from each truck type to derive the solid waste quantities.

If landfill records are either unreliable or nonexistent, a landfill survey may be required. Landfill survey methods involving weekly or seasonal truck counts and weighing programs have been demonstrated (2,6). These data gathering programs can be quite expensive to conduct depending on how much data are gathered. The investigator must decide in advance what level of data precision is needed before embarking on a landfill survey program. Some variables to consider are:

Survey Period - Survey data can be taken for a day, week, month, or for several week-long periods during different parts of the year, to establish seasonal variation. A minimum of one week is suggested to establish data which is at least minimally representative of the total waste stream.

Number of Landfills Surveyed - If more than one landfill is utilized by the study area, it may be necessary to conduct a surveillance at each. An alternative is to gather survey data at one landfill which has a strictly-defined service area for the purpose of establishing unit waste factors for the entire study area. However, if only one landfill is used to establish unit waste factors, caution must be exercised to ensure that data from this landfill is representative of the total study area generation.

Type of Data - The type of data taken can be limited to a record for each vehicle entering the landfill which includes the type of vehicle (packer, frontload packer, transfer trailer, pick-up), vehicle size (dimensions or payload cubic yards), the hauler (contractor, city, private citizen), and an estimate of waste category (residential, commercial, industrial, commercial-industrial mixed). Other data which may be taken include time of arrival, and vehicle weight (by using portable scales).

The reader should consult the reference materials (2,6) for further explanation of survey methods.

The data from either of these waste quantity estimating methods (scale data, volume or load data, or landfill survey) should result in an estimate of the number of tons disposed by the study area in the current year, and if possible a breakdown of quantities disposed by month. It is also helpful to compile comparable data for the past three to five years so that growth trends can be observed.

3. Existing Composition

Determining the feasibility of resource recovery as a solid waste management option requires knowledge not only of the quantities of solid waste generated, but also of the composition of the waste. Knowledge of composition in two general areas is necessary: waste composition by material type, and waste combustion characteristics. Waste composition is the basis for calculating the quantities of recoverable materials as well as the potential energy yield from the solid waste.

The most direct way to determine the local composition of the processable waste stream is to collect representative samples for composition analysis at the landfills, transfer stations, or incinerators currently utilized by the study area. Several references discuss the methods which can be used to obtain representative samples for laboratory analysis (6-8). These methods require significant expenditures of time and money, but are necessary in order to establish final design data for a resource recovery facility. The reader is directed to the reference materials for details.

Since this book deals with a level of detail required for initial feasibility analysis, an alternative to the direct landfill survey will be presented. The alternative is to examine the results of landfill sampling analyses in other areas of the country along with other

published data to form a judgment of the likely composition of the
waste in the study area. Although not as accurate as a landfill sample
analysis, a reasonably accurate composition can be selected from these
data sources. At a minimum, the percent (by weight) of all
combustibles, ferrous metals, aluminum, and all other inorganics should
be established for material composition. Combustibles and other
inorganics can be separated into narrower categories if recovery of
these materials is anticipated. Combustible categories are paper,
cardboard, yard waste, plastic, etc. Categories for inorganics might
include glass, mixed non-ferrous metals, and miscellaneous. The
combustion characteristics which should be established are: as-received
heating value in Btu/lb, moisture content in percent by weight and ash
content in percent by weight. Material composition and combustion
characteristics should be established for raw, as-received processable
solid waste.

The U.S. Environmental Protection Agency (EPA) has made
estimates on a national basis of the composition of "post-consumer"
(residential and commercial) solid waste (9). In addition, Table 2.2
displays the composition data developed from landfill sample analyses in
several parts of the country for residential and commercial wastes.
Other data sources for combustion characteristics include data published
by EPA for solid waste at St. Louis (10), and Ames, Iowa (11), as well
as other solid waste literature (1,12,13). Both the national level data
and these other local sample analyses should be applied with caution to
any particular local-level planning effort. Local factors to consider in
selecting both material composition and combustion characteristics
include:

Weather - The amount of annual rainfall can affect the moisture
content and heating value.

Local Source Separation - In states which have mandatory beverage
container deposit laws, ferrous metals and aluminum percent
compositions may be lower. In cities which conduct regular separate
collection of newsprint, the combustibles category may need to be
reduced.

Geographic Location - In remote areas where delivery of fresh food is minimal, packaging wastes such as steel and aluminum cans may occur in higher percentages.

In general, however, an examination of Table 2.2 reveals a surprising uniformity of waste composition in disparate parts of the country.

4. Unit Waste Factors

Given the existing quantities, a resource recovery analysis must proceed with a projection of those quantities over a selected planning horizon. It is likely that as a community grows, waste quantities will also grow. An extention of past waste quantity growth trends from historical landfill data is one way to approach making predictions of future quantities, but one which does not take into account factors which might modify past growth rates.

Another way to make waste quantity projections is to relate the generation of waste to characteristics of the study area which are easily determined, and for which there are widely accepted future projections. Unit waste factors can be calculated and used along with a projection of the community characteristic to derive future waste quantities. Since solid waste is a consequence of the activities of people, an obvious choice for this community characteristic is population. The use of population-based unit waste factors (pounds per capita per day) is a widely used and accepted practice for projecting solid waste quantities. For certain types of wastes (commercial and industrial), the number of employees in certain business categories has also been used. There have been attempts to relate solid waste generation to other factors such as income level, retail sales volume, and others (14), but population and employment have consistently been shown to be the best predictors of solid waste generation and continue to be the most easily obtainable data in projected form. These data are usually available from city or county planning departments, transportation departments or the regional planning agency with jurisdiction in the study area.

TABLE 2.2
Solid Waste Composition Analyses

SOLID WASTE COMPONENT	IOWA		MINNESOTA			CALIFORNIA				MONTANA			
	DUBUQUE (RES)	DUBUQUE (COMM)	ST CLOUD (RES)	ST CLOUD (COMM)	OLMSTEAD CO (RES-COMM)	COLTON (RES)	COLTON (COMM)	SAN DIEGO (RES)	SAN DIEGO (COMM)	MISSOULA	BUTTE	BILLINGS	GREAT FALLS
PAPER	37.0	42.2	37.0	36.1	33.4	26.9	35.4	38.6	44.1	25.0	24.3	24.9	26.9
CARDBOARD	3.5	11.0	14.0	22.6	12.8	6.2	20.4	6.8	22.8	10.3	7.0	10.1	8.2
PLASTIC	5.3	7.8	4.1	3.7	5.6	2.8	4.5	3.6	7.5	4.3	6.1	6.1	4.2
WOOD	0.6	1.0	2.3	1.6	2.0	2.2	4.5	1.4	3.9	2.2	0.1	1.0	1.5
FOOD WASTE	10.6	7.4	17.5	1.7	14.6	3.4	2.6	2.8	5.5	12.9	21.9	20.5	13.6
YARD WASTE	25.1	7.2	0.6	0	9.1	40.8	13.6	33.7	2.3	29.6	14.3	12.2	28.0
TEXTILES	2.3	1.7	3.6	4.4	3.2	2.5	6.3	2.3	2.6	3.2	3.9	6.0	2.7
RUBBER (LEATHER)	0.2	0	1.0	1.6		0.9	1.3	1.1	0.7				
RESIDUE			2.6		1.7								
TOTAL PERCENT COMBUSTIBLE	84.6	78.3	82.7	85.0	80.7	85.7	88.6	90.3	89.4	87.5	77.6	80.8	85.1
FERROUS	8.8	13.6	8.0	8.8	9.5	5.5	5.6	4.5	5.2	6.2	9.0	9.0	6.7
ALUMINUM	1.1	1.1	0.5	0.3	0.8	0.6	0.6	1.0	0.8	1.4	2.5	1.8	1.7
GLASS	5.4	6.7	8.8	5.9	9.0	5.5	2.9	4.0	4.3	4.9	10.9	8.4	6.5
RESIDUE	0.1	0.3				2.7	2.3	0.2	0.3				
TOTAL PERCENT NON COMBUSTIBLE	15.4	21.7	17.3	15.0	19.3	14.3	11.4	9.7	10.6	12.5	22.4	19.2	14.9
BTU/lb (AS RECEIVED)	3653	4796	3793	4155		4878.00		6456.00		4843	6049	4519	4748
BTU/lb (DRY)	7010	8173								7746	7402	7739	7278
BTU/lb (AVERAGE)	3600	5300	4000										
% MOISTURE	41.1	36.6	39.4	33.6		28.0		21.9	20.9	37.8	26.6	41.3	34.9
% RESIDUE	15.1	8.7	14.1	18.9						13.3	8.7	11.3	11.9
CARBON	29.1	40.8	23.9	29.9						43.7	45.7	43.0	41.5
HYDROGEN	2.3	2.2	5.1	3.3						6.2	6.6	6.2	5.6
OXYGEN	11.6	11.2	16.5	16.7						35.2	39.1	37.5	39.6
NITROGEN	0.52	0.37	0.64	0.56						0.88	1.01	1.07	0.68
CHLORINE	0.17	0.15	0.25	1.47						0.60	0.49	0.75	0.45
SULFUR	0.02	0.02	0.12	0.53						0.12	0.11	0.31	0.23

AVG. 5039

TABLE 2.2
Solid Waste Composition Analyses

SOLID WASTE COMPONENT	MICHIGAN MARQUETTE (RES-COMM)	ARIZONA PHOENIX (RES)	ARIZONA PHOENIX (COMM)	GEORGIA DEKALB (RES)	GEORGIA DEKALB (COMM)	FLORIDA ST PETERSBURG (RES)	ILLINOIS SPRINGFIELD (RES)	ILLINOIS SPRINGFIELD (COMM)	ILLINOIS SPRINGFIELD (RES-COMM)	WISCONSIN REGION I (RES)	WISCONSIN REGION I (COMM)	NATIONAL NCRR (RES-COMM)	NATIONAL EPA 4TH REPT
PAPER	⎱46.6	⎱43.7	⎱50.8	⎱37.3	⎱58.2	⎱31.4	27.6	21.7	25.9	25.4	27.4	⎱42.7	⎱35.0
CARDBOARD	⎰	⎰	⎰	⎰	⎰	⎰	4.2	22.7	9.4	10.2	36.1	⎰	⎰
PLASTIC	7.0	4.1	5.3	3.5	4.5	1.3	5.3	5.1	5.3	3.2	3.3	1.7	3.8
WOOD	0.8	1.3	2.3	3.9	2.5	1.9	1.7	3.9	2.3	5.3	10.0	2.5	3.8
FOOD WASTE	13.8	12.2	12.5	26.6	2.7	0.8	15.5	18.6	16.4	17.2	11.0	14.6	14.9
YARD WASTE	10.0	17.2	6.9	3.2	0.5	46.7	21.0	2.3	15.8	24.1	—	12.5	16.3
TEXTILES	3.2	3.8	2.5	0.7	3.3	2.9	3.9	1.5	3.2	⎱2.1	⎱0.9	2.4	1.7
RUBBER (LEATHER)				13.2	0.6	0	0.4	0.1	0.3	⎰	⎰	1.8	2.6
RESIDUE	5.7	5.7	6.5		10.7		4.2	1.5	3.3	0.3	3.6		
TOTAL: PERCENT COMBUSTIBLE	81.4	88.0	86.8	89.7	83.0	85.0	83.8	77.4	81.9	87.8	92.3	78.2	78.0
FERROUS	8.1	4.9	5.6	5.5	10.7	5.4	7.1	12.4	8.6	6.4	5.4	8.2	⎱9.8
ALUMINUM	1.3	0.9	0.5	1.0	1.0	1.0	0.7	0.6	0.7	1.1	0.1	0.9	⎰
GLASS	8.7	6.2	7.1	3.8	5.3	5.7	5.8	8.3	6.5	4.7	2.0	10.3	10.5
RESIDUE	0.5					2.9	2.6	1.3	2.3		0.2	2.4	1.6
TOTAL PERCENT NON COMBUSTIBLE	18.6	12.0	13.2	10.3	17.0	15.0	16.2	22.6	18.1	12.2	7.7	21.8	21.9
BTU/lb (AS RECEIVED)	5000	5000		4810	5227		5470.0	4972.9	5330.8				
BTU/lb (DRY)							7680.3	7953.6	7756.8				
BTU/lb (AVERAGE)									5331				
% MOISTURE		29.1	32.5	9.7			28.6	37.5	31.1				
% RESIDUE				37.1	32.1		12.2	10.3	11.6				
CARBON							42.8	44.3	43.2				
HYDROGEN							5.2	5.1	5.2				
OXYGEN							39.6	39.4	39.6				
NITROGEN							0.6	0.6	0.6				
CHLORINE				0.04			0.09	0.06	0.08				
SULFUR				0.09			0.08	0.18	0.11				

There are different approaches to establishing unit waste factors for a particular study area. The choice of a particular approach depends on the level of accuracy required for the projections and the kinds of waste quantity and demographic data available for the study area. These different approaches are detailed in the following paragraphs.

a. Composite Unit Waste Factor - The composite unit waste factor is the simplest to establish for a particular study area. This "unit waste factor" is total processable solid waste (residential, commercial, industrial) divided by total population. The investigator should confirm that the waste quantity used in this calculation is 1) processable waste only, and 2) representative of the waste generated in only the study area. These adjustments to landfill data should be made before calculating the unit waste factor. Also, the investigator should confirm that the population data used in the calculation is representative of the same year as the solid waste quantity data.

In order to avoid a seasonal bias in the unit waste factor, an annual waste generation quantity should be used in the calculation, rather than a monthly, or daily generation quantity. This calculation will result in a unit waste factor with units of tons per capita per year which can be converted to pounds per capita per day by multiplying by a factor of 5.479 (2,000 divided by 365).

The magnitude of the composite unit waste factor will depend on the amount of commercial and industrial activity in the study area as well as individual residential waste practices. Table 2.3 gives composite rates calculated from scale data in several areas around the country along with the EPA national average (9) for comparison purposes. If landfill records are not available for the study area, and if it is not possible to conduct a landfill survey because of time and/or monetary constraints, a composite unit waste factor can be established by examining past studies such as these. However, because of wide variability in unit waste factors, it is unwise simply to calculate an

TABLE 2.3

Composite Unit Waste Factors for
Processable Solid Waste

Location	Year	Composite Unit Waste Factor (lb/capita/day)	Reference
Fairfax Co., VA	1978	3.95	(15)
Omaha - Council Bluffs Region, (Nebraska,Iowa)	1978	3.27	(2)
Pinellas Co., Florida	1975	4.3	(16)
Phoenix, Arizona	1978	6.23	(5)
Charlotte, N.C.	1977	6.7	(17)
EPA, National Average	1975	3.2	(9)

average from a table such as Table 2.3. It is better to select the factor calculated for a comparable locality. The best approach, however is to utilize the most recent local data to establish this factor rather than relying on factors calculated for either study areas.

b. Separate Unit Waste Factors - The use of a composite unit waste factor is simple, but it disguises certain useful information. Recognizing that different kinds of waste may be generated in different sections of the study area (i.e., residential waste in residential zones, commercial waste in downtown central business districts, etc.), separate unit waste factors can be established for different kinds of waste. By using separate unit waste factors along with population and employment information for different areal districts within the study area, a more detailed account of waste generation can be obtained.

The use of separate unit waste factors is especially important when the study area, and related districts established for the systems analysis (see Chapter 6) are geographically small. For example, if a downtown central business district (or any other primarily commercial or industrial area) is one district, a population-based composite unit waste factor might assign a small waste generation amount for this district because of the small resident population. Alternatively, if separate unit waste factors were established with an employee-based commercial unit waste factor, a more accurate prediction of waste generation for this district would be made.

Separate unit waste factors cannot be established, however, unless certain data are available. If landfill data are unavailable, and a landfill survey cannot be performed because of time or monetary constraints, an attempt to establish separate unit waste factors from other studies is not advisable. A composite factor should be used in this case. Also, landfill data must be broken down into residential, commercial, and industrial categories, or at a minimum into residential and commercial-industrial categories. If not already available in these categories, many times it is possible to examine the landfill hauler records for a breakdown if it is known which kinds of wastes are collected by each hauler. In addition, before calculating separate unit waste factors, the investigator should confirm that both population and

TABLE 2.4

Residential Unit Waste Factors

Location	Year	Residential Unit Waste Factor (lb/capita/day)	Reference
Lake County, Illinois	1971	2.3	(18)
Wayne County, Michigan	1976	2.52	(19)
Wyandotte, Michigan	1976	2.38	(20)
Grand Rapids, Michigan	1975	2.3	(21)
State of Montana	1975	2.3	(22)
Dubuque, Iowa	1975	2.2	(23)
Pinellas County, Florida	1976	2.0	(16)
Phoenix, Arizona	1978	3.3	(24)
Dallas-Ft. Worth Texas	1975	2.75 to 3.23	(25)
Omaha-Council Bluffs, Nebraska-Iowa	1979	1.8	(26)
Charlotte, North Carolina	1977	2.4	(17)

employment data are available in projected form for the planning horizon desired and for acceptable areal districts. If the above data are available, the following types of separate unit waste factors can be established;

Residential - This unit waste factor is calculated by dividing the total annual tonnage of residential waste delivered to area landfills by the total population for the study area. Conversion to units of pounds per capita per day can then be made. The investigator should confirm that the waste tonnage represents mostly (if not all) residential type waste and that the year represented by the solid waste quantity and the population data is the same. Past studies of residential unit waste factors based on scale data for regions near the study area should also be compared. Table 2.4 displays residential unit waste factors calculated from landfill scale data for different areas of the country.

Commercial - It has been shown that solid wastes generated by commercial establishments can be mathematically related to characteristics of the establishments in question. Commercial solid waste generation is most closely related to the number of employees and the type of business involved (27). In general, commercial solid waste is generated by businesses involved in Wholesale Trade (SIC 50-51), Retail Trade (SIC 52-59), Finance, Insurance, Real Estate (SIC 60-67), Services (SIC 70-89), and Government (SIC 91-97). Landfill scale or volume data for wastes from these establishments should be used in this calculation. The annual quantity in tons of commercial solid waste is divided by commercial employment for the study area and converted to pounds per employee per day to establish the commercial unit waste factor. Attempts have been made to establish separate commercial unit waste factors for different types of businesses (27-29); however, this level of detail is not required for an initial feasibility study. Employment data are usually available from city or county planning departments, transportation departments, or the regional planning agency with jurisdiction in the study area.

TABLE 2.5

Commercial Unit Waste Factors

Location	Year	Commercial Unit Waste Factor (lb/emp./day)	Reference
Roscommon County, Michigan	1975	5.7[a]	(29)
Southeastern Michigan Council of Governments (Detroit area)	1973	13.4[a]	(28)
North-Central Texas Council of Governments (Dallas-Ft.Worth area)	1974	8.3	(25)
Hennepin County (Minneapolis-St.Paul), Minnesota	1975	3.21	(30)

[a] Weighted averages of unit waste factors for separate business types based on employment data for Wayne County Michigan.

TABLE 2.6

Industrial Unit Waste Factors[a]

Location	Year	Industrial Unit Waste Factor (lb/emp./day)	Reference
Racine County, Wisconsin	1977	6.0[c]	(32)
Arlington, Texas	1974	6.62[b]	(25)
Dallas, Texas	1974	14.0[b]	(25)
Hennepin County (Minneapolis), Minnesota	1975	7.92[b]	(30)
Southeastern Michigan Council of Governments (Detroit area)	1973	10.4 to 12.4[c]	(28)
State of Michigan	1972	23.8[c]	(31)

[a] Does not include "special" industrial wastes such as toxic chemicals, oils, tars, solvents, ash, foundry sand, slag, scale, etc.
[b] Based on scale and route survey data.
[c] Based on questionnaire survey.

Certain state employment agencies also maintain employee data for counties, and sometimes for cities or "labor market areas." Studies of commercial unit waste factors in other similar study areas should be examined for comparison. Table 2.5 lists commercial unit waste factors calculated in several study areas around the country. The wide variation in the data given in Table 2.5 underscores the importance of developing commercial unit waste factors from reliable local information.

Industrial - Studies in several areas of the country have shown that industries usually generate more processable waste per employee than commercial businesses. Industrial businesses are generally those in the Standard Industrial Classification (SIC) code groups 19 through 39 (Manufacturing) and 40 through 49 (Transportation, Communication and Utilities). The reasons for this phenomenon are not totally understood, but the data given in Table 2.6 demonstrates this fact. Note that the unit waste factors on Table 2.6 do not include "special" industrial wastes such as hazardous wastes, ash, foundry sand, slag, sludge or other such materials. This unit waste factor should be calculated in the same manner as the commercial unit waste factor utilizing landfill records of waste delivered from industrial sources and industrial employment data from the study area. Industrial unit waste factors have also been established by utilizing a questionnaire survey of industries in the study area (28,31,32). This method might be used in lieu of landfill data, but questionnaire survey data reliability is not high and it is often difficult to get a high enough response rate to develop a representative sample.

Commercial/Industrial - Many times it is impossible or impractical to separate commercial and industrial waste quantities in the available landfill data. This problem arises when private collectors serve both commercial and industrial accounts on the same route which results in mixed loads. In this situation, a combined commercial/industrial unit waste factor can be developed with satisfactory accuracy. The

TABLE 2.7

Commercial/Industrial Unit Waste Factors

Location	Year	Commercial/Industrial Unit Waste Factor (lb/emp./day)	Reference
Charlotte, North Carolina	1977	8.4	(17)
Omaha-Council Bluffs,Neb.& Iowa	1978	3.3	(26)
Des Moines, Iowa	1968	8.0	(33)
Quad Cities, Illinois, Iowa	1972	10.2	(34)
Phoenix, AZ	1975 1978	5.8 7.6	(35)
DeKalb County (Atlanta), Georgia	1975	6.3	(36)
Hennepin County (Minneapolis), Minnesota	1975	11.1	(30)
Dallas-Ft. Worth, Texas	1974	10.0	(25)

combined commercial and industrial waste quantity is divided by total employment in the study area, then converted to pounds per employee per day as previously discussed. It may also be necessary to utilize a combined commercial/industrial unit waste factor in a study area for which only total employment (rather than separate employment figures for commercial and industrial businesses) is available in projected form by areal district. Table 2.7 displays combined commercial/industrial unit waste factors calculated from scale data or landfill surveillance data for several study areas around the country.

5. Unit Waste Factor Projection

When projecting solid waste quantities, it is necessary to know how the unit waste factors may change in the future in addition to population and employment changes. Historically, the national composite unit waste factor has been steadily increasing. It has been estimated that total processable amounts collected in urban areas have risen from 2.75 pounds/capita/day in 1920 to 5.1 pounds/capita/day in 1970 (1). Based on these historical trends and predictions of increases in per capita consumption of consumer products, predictions of per capita waste generation have been made (37).

Waste generation data after 1973 in several areas of the country and on a nationwide basis indicate, however, that this trend has been interrupted. Landfill scale data from Phoenix, Arizona, (5) show steadily increasing quantities disposed until 1972 when the tonnage peaked. Quantities decreased dramatically in 1973 and continued to decrease through 1975 despite rising population and employment over the same period. The composite unit waste factor for Phoenix has, however, risen from 5.43 pounds/capita/day in 1975 to 6.23 pounds/capita/day in 1978. Data for Southgate, Michigan have shown a similar decrease in the composite unit waste factor after 1973 (20). Landfill scale data for Fairfax County, Virginia, (4) also confirm a steady decline in composite unit waste factor from 1974 through 1978. Table 2.8 displays these data. Further, national data for "post-consumer" (primarily residential and commercial) solid waste indicate a

TABLE 2.8

Composite Unit Waste Factors 1974-1978
Fairfax County, Virginia

Year	Population	RCI Solid Waste (tons/yr)[a]	Composite Unit Waste Factor (lb/capita/day)[b]
1974	522,200	425,878	4.47
1975	537,200	422,749	4.31
1976	554,500	429,862	4.25
1977	567,600	417,244	4.03
1978	578,900	417,581	3.95

[a] Reference 4
[b] RCI = residential, commercial, and industrial wastes combined from County landfill scale records

similar, post-1973 decline in the composite unit waste factor (9). These
EPA figures show a decline in total waste disposal from 135 million
tons in 1973-74 to about 128 million tons in 1975 with a composite unit
waste factor drop to 3.2 pounds/capita/day in 1975 from 3.5
pounds/capita/day in 1973.

The reasons for this decline are not completely understood. The
general economic recession beginning in mid-1974 certainly appears to
have influenced waste generation. A shift in consumer attitudes about
excess packaging and disposable products may have also been a
contributing factor. Further, it is not clear whether this decrease was
a short-term change or an indication of a long-term trend in solid
waste generation. The Phoenix data suggest that this trend has
reversed, but the Fairfax County data suggest a continued decline.
These conflicting data make even more difficult the speculative task of
projecting future unit waste factors.

In selecting future unit waste factors for a particular study area,
the investigator should consider that it is generally more desirable to
underestimate waste quantities than to overestimate them in a resource
recovery feasibility analysis. An underestimate means that if resource
recovery is chosen, longer operating hours will be required, or in the
case of a facility with a market-limited capacity, more waste than
projected will be landfilled. An overestimate will result in building a
facility with more capacity than is necessary unless the capacity is
energy market-limited. Because of the high fixed costs (debt service
and others) characteristic of these facilities, an over-sized facility can
result in per-ton costs significantly higher than expected.
Conservatively low unit waste factor projections, or the use of the
same unit waste factor throughout the study period, is therefore
advisable.

6. Future Quantities

The calculation of existing and projected unit waste factors has been
previously discussed. In order to project the tonnages of solid waste

for a study area, the investigator must obtain population and in some cases employment projections for the study area for use along with the unit waste factors.

A projection period must first be selected. It is good practice to select a projection period which matches the expected financing period for a potential resource recovery facility. A popular convention in municipal bond financing is twenty years. Since a period of two to three years is many times required between the completion of a feasibility study and the bond sale, it is wise to select a period of 22 to 25 years.

Since a subsequent part of the feasibility analysis is concerned with optimizing the location of solid waste facilities (landfills, transfer stations, and resource recovery facilities), it is necessary to divide the study area into areal districts and make quantity projections for each. It is therefore not sufficient to obtain only a projection of total population and/or employment for the study area. Time and distance information between each zone is also required for the waste transport analysis. Many demographic and transportation planning models utilize the census tract as the basic areal unit. Census tracts are, however, too small for this type of analysis because the large number of calculations which would be required by using this small areal unit are not justified. It is also common to find that city and county planning departments, or regional planning agencies have aggregated census tract information into more manageable "planning zones," or "transportation analysis zones." The use of these larger areal districts for waste generation districts (WGD) will normally provide more than sufficient detail for the systems analysis. It may be advisable to further aggregate the transportation analysis zones in order to further simplify the analysis. In general, the size of each waste generation zone should be such that no more than three to four percent of the total waste generation is composed by each.

Depending on the use of either composite or separate unit waste factors, employment, and/or population should be tabulated for each WGD in (at most) five-year increments. Demographic projections are normally made on a five-year increment, but some are on a ten-year basis.   If five-year increment data are not available, the interstitial years can be calculated by linear interpolation without the loss of significant accuracy.   The unit waste factors are then multiplied by the zonal employment and/or projections for each zone.   Separate projections of residential, commercial, industrial, or commercial/industrial waste can then be summed for each district to derive the total processable waste generation by WGD.   Summing the WGD totals will result in the total processable waste generated in the study area for each projected year.   The non-processable waste quantity which must be a factor in landfill sizing can be calculated by utilizing the proportion of total waste which is in the non-processable category established from existing landfill records.   This same proportion can be used in all projected years to establish an approximation of the non-processable waste quantity.

7. Future Composition

It is clear that the composition as well as the quantity of processable solid waste shifts over time.   Prediction of the direction and magnitude of these shifts is very difficult.   A high degree of uncertainty in these predictions exists because the factors which influence changes in composition and the vastly complex interactions between these factors are not well understood.

Attempts have been made by the EPA and others to project changes in solid waste composition to 1990 (38,39) based upon expected changes in consumption patterns.   Most of the changes calculated in these references are in the plus or minus ten to fifteen percent range. The most important changes are increases in percent composition in all paper categories, a decrease in glass composition, a slight increase in ferrous metal composition, and an increase in aluminum composition.   In general, these projected shifts will benefit an energy and materials recovery facility.

Given the uncertainty under which these predictions of general composition shifts were made, and considering the relatively small magnitude of the changes, it might be logical to assume that the existing composition found in the study area under examination will remain relatively constant. These national predictions were, however, made in the absence of information about local source reduction and source separation programs which may significantly change processable solid waste composition in certain categories. Although such programs are an important part of any comprehensive solid waste management plan, a complete discussion of source reduction and source separation programs is beyond the scope of this book. The effect of certain types of these programs on local processable waste composition, must however, be considered.

A source reduction method which is gaining acceptance in several states and some counties is mandatory beverage container deposit legislation. Oregon, Vermont, Maine, Iowa, and Michigan all have similar state legislation. Experience in Oregon (40) has shown a significant reduction in the number of discarded beverage containers.

A national mandatory deposit law has also been under consideration for several years. Federally-sponsored research (41-43) has estimated that under a federal mandatory deposit system, a 70 percent reduction in the beverage container component of solid waste would occur. National composition data (9) indicate that glass beverage containers (excluding wine, liquor, and other food containers) constitute 46.0 percent of all glass in the waste stream; aluminum beverage containers constitute 48.8 percent of all aluminum in the waste stream; and steel beverage containers are 11.7 percent of all ferrous metals in the waste stream. Using a 70 percent reduction in each material category, a processable waste tonnage reduction of about 4 percent would be expected.

A study sponsored by the Federal Energy Administration (FEA) (44) estimated that under a federal mandatory deposit system, annual production of glass beverage containers (both refillable and non-refillable) would decrease by about 9 percent while production of steel and aluminum beverage containers would decrease by about 50 percent and 80 percent respectively. These reductions would have a processable solid waste reduction impact of about 1 percent of the total waste stream.

The EPA has also generated what it feels to be the maximum expected recovery rates from various types of local source separation programs under ideal conditions (45). If these maximum rates are used along with the mandatory beverage container reduction estimates by the FEA previously presented, the composition shifts given in Table 2.9 would be expected based on the national composition for 1975 (9). Note that under this hypothetical scenario, total combustibles are reduced by about 7 percent and total waste is reduced by about 6 percent.

When adjusting existing composition to account for future source separation and/or source reduction programs, the investigator should note that the reduction percentages given previously are maximums based on ideal conditions and participation rates higher than have normally been observed in past programs. Also, local estimates of the percent of each material category affected by the particular local program going into effect should be made.

B. Haul Costs

Since a later part of this analysis (Chapter 6) is concerned with optimizing facility locations in the study area, it is necessary to know the costs of moving the processable solid waste. In this optimization, the investigator should be concerned only with the costs of moving the waste after collection from the end of the collection route. Although collection costs are a major component in the total costs of solid waste management, the collection system and its associated costs will likely remain essentially the same for all resource recovery and landfilling alternatives. Collection costs are therefore removed from the analysis.

Unit haul costs (dollars/hr or dollars/mile) should be calculated for moving waste in three different modes: 1) hauling waste in the collection vehicle, 2) hauling waste in a larger transfer vehicle, and 3) hauling residue and ash in a suitable vehicle from the resource recovery facility. Methods for developing unit haul costs for these three modes are discussed in the following paragraphs.

TABLE 2.9

Maximum Estimated Composition Shifts Due To Source Separation and Source Reduction Programs

Material Category	National 1975 % Composition[a]	% Affected by SS and SR Programs[a]	% Maximum Achievable Reduction[b]	New Composition lb/100 lb[c]	New Composition %
1. Total Paper	34.9	16	-	29.65	31.7
Newsprint	5	100	50	2.50	2.7
Corrugated	7	100	25	5.25	5.6
Office Paper	4	100	25	3.00	3.2
Other Paper	18.9	0	0	18.90	20.2
2. Other Organics	43.1	0	0	43.10	46.1
3. Total Combustible[d]	78.0	20.5	-	72.75	77.8
4. Glass	10.5	46	9	10.07	10.8
5. Ferrous Metals	8.9[e]	11.7	50	8.37	8.9
6. Aluminum	0.7	48.8	80	0.43	0.5
7. Other nonferrous	0.3	0	0	0.30	0.3
8. Misc. inorganics	1.6	0	0	1.60	1.7
TOTAL	100.0	--	--	93.52	100.0

a Reference (9), based on "as-disposed" data.
b References (44) and (45).
c Pounds of new composition per 100 pounds of original composition.
d Sum of Categories 1 and 2.
e Estimated "as-disposed" percentage.

1. Collection Vehicle

Hauling waste in the collection vehicle after collection is commonly
called the "primary haul" because it is the first movement of the waste
(see Figure 2.1).   Examples of primary hauls are the movement of
waste from the end of the collection route to a landfill, transfer
station, or resource recovery facility.

The costs of transporting waste in the collection vehicle can be
divided into two categories: vehicle costs and labor.   Vehicle costs
should include fuel, maintenance, insurance, oil, tires, and depreciation.
These cost data for collection vehicles should be available from city or
county accounting records for municipal collection vehicles, and from
the private hauler records if private collection is involved.   These
vehicle costs are most conveniently represented on a dollar per mile
basis.   Note however, that it is not sufficient to merely divide a
collection vehicle's annual operating costs by its annual odometer
mileage.   This calculation gives an excessively high per-mile operating
cost for the transporting of solid waste after collection because the
total annual costs include the costs of the vehicle in the collection
phase as well as the transporting phase.   Fuel and maintenance costs
per mile are much higher during collection than during the transporting
phase because of frequent starts and stops and operation of the
compaction equipment.   Therefore, the vehicle cost accounting records
should be examined to select maintenance costs for the chassis and the
cab only (exclude maintenance costs for the packer equipment),
depreciation on the chassis and cab only, oil and tires.   Fuel costs
should be derived by determining from the operators the average miles
per gallon obtained during transportation after the vehicle is fully
loaded, then multiplying by the cost per gallon of fuel.

Labor costs for the collection vehicle should include driver and
collection crew salaries (if the crew accompanies the collection vehicle
during transportation, which is normally the case).   The investigator
should also be certain that the costs for fringe benefits, overhead, and
management costs, in addition to direct salaries, are also included.

These unit costs are most conveniently expressed in units of dollars per hour. These data are normally available from city salary schedules and (in the case of private haulers) from private hauler records. These dollars/mile and dollars/hour unit costs can then be used along with time and distance data in the systems analysis to optimize facility locations.

Since the type of vehicle, the size of the collection crew, and the type of hauler (municipal or private) may vary from WGD to WGD, these unit costs may be assigned to each WGD individually by waste type. Using this method, it is also necessary to assign a payload capacity by waste type for each vehicle in tons for each WGD. For example, a particular WGD may generate residential, commercial, and industrial processable waste, each hauled by different trucks and crews. Unit haul costs and payloads for each waste type could be established for this WGD. An alternative to assigning individual unit costs and payloads is to develop average unit costs and payloads for use in groups of districts or for the entire study area. These averages are sometimes called "composite vehicle" data. When vehicle usage varies within a WGD, data for a "composite vehicle" must be calculated for this WGD. Calculating total study area haul costs utilizing individual unit haul costs for each waste type and WGD is possible theoretically, but as a practical matter is cumbersome and time-consuming. The method which yields reasonable accuracy and consumes a minimum of calculation time is to establish "composite vehicle" data for each WGD. This subject is discussed more thoroughly in Chapter 6.

2. Transfer Vehicle

Transfer vehicles are the larger vehicles (65 to 112 cubic yards) utilized by a transfer station for long transportation distances. Transfer vehicles may also be used to transport prepared refuse-derived fuel from a processing plant to the fuel user. Transportation in these vehicles is sometimes called a "secondary haul."

Although barge and rail vehicles have been used for hauling solid waste, most applications require the use of road vehicles (tractor-trailers) for transfer. Generally, there are two types of transfer trailers, "open-top" and "compacted." The open-top vehicle size is generally between 90 and 100 cubic yards, and can achieve compaction factors of about 400 lb/CY (pounds per cubic yard) by direct dumping from a tipping floor with leveling and tamping by a hydraulic backhoe arm. The compacted transfer trailer is smaller (generally 65 to 75 cubic yards) and can achieve compaction factors of about 500 lb/CY with a stationary compactor located at the transfer station. Table 2.10 lists size and payload capacity data for both open-top and compacted trailer designs in use in several areas of the country.

Since different payloads and unit costs are associated with these different transfer trailer designs, the investigator must select a design for use in the analysis. (See Chapter 4 for a detailed discussion of transfer station design.) If an existing transfer station is operating in the study area, and cost data are available, it is advisable to use these data. If no transfer station exists in the study area, a selection can be made (note that this selection affects not only vehicle costs, but transfer station design and its associated costs as well).

Regardless of the design chosen, vehicle costs should include depreciation, fuel, oil, tires, maintenance and insurance. Since this vehicle is not used in a collection function, costs for both the transfer trailer and tractor (cab portion) should be included. Labor costs should be handled in a different manner for transfer vehicles than for collection vehicles. Since every hour spent in transporting waste to a landfill, transfer station, or resource recovery facility in a collection vehicle is time taken away from the collection function, labor costs for the driver and crew should be included in transport costs for a collection vehicle. A transfer trailer driver, however, does not perform the collection function, therefore the driver's utilization does not vary with the time actually spent on the road. That is, the transfer vehicle driver is paid a certain salary no matter how many miles he drives in a month. It is therefore advisable to account for driver labor in the costs of operating the transfer station rather than assigning an hourly cost to the transfer vehicle.

TABLE 2.10

Transfer Trailer Size and Density Data

Location	Trailer Type	Trailer Size (CY)	Average Payload (tons)	Average Compaction Factor (lb./CY)	Comment	Reference
Los Angeles, CA	open-top	112	22	393	Direct Dump, Tamped	(47)
Harvey, ILL	open-top	102	18.75	367	Direct Dump No Tamping	(48)
Seattle, WA	open-top	92	20	435	Storage Pit with Crawler Tractor, Tamped	(49)
Bel-Aire, CA	open-top	112	23.5	420		(50)
Wyandotte, MI	compacted	75	17.5-25[a]	467-667[a]	Stationary Compactor	(51)
Dallas, TX	compacted	65	16.25	500	Stationary Compactor	(3)

[a] Data range, averages not available.

50

3. Residue and Ash Vehicle

An essential cost of operating a resource recovery facility, and a cost
which will influence the choice of its location, is the cost of hauling
residue and/or ash from the facility to a landfill. This hauling function
is also usually classified as a "secondary haul." The quantity of these
materials can range from 20 percent to 35 percent of the incoming raw
waste by weight (see Table 4.4, Chapter 4). Generally a 30 to 40
cubic yard dump truck (covered in the case of ash hauling) can be used
for this function. Vehicle costs should be calculated in a manner
similar to the method previously discussed for the transfer vehicle.
Labor costs should generally be included as a per-hour cost for hauling
because the driver will usually have other duties in plant operation, just
as collection crew labor is included as a per-hour cost for hauling.

C. Potential Facility Sites

An essential part of the basic data gathering phase of a resource
recovery feasibility analysis is to establish potential sites for landfills,
transfer stations, and various types of resource recovery facilities. A
final selection of a single particular site is not the purpose of this
task. It is, however, necessary to select a set of sites for each
particular facility which meet certain minimum physical and locational
criteria for use as candidate sites in the systems analysis. The systems
analysis will display the relative haul costs associated with the use of
each site. A final selection is normally made after consideration of
social and political issues. Unfortunately, many times a site is selected
not on the basis of engineering criteria, but as a result of political
trade-offs.

The following paragraphs describe the methods for making the
selection of an initial set of candidate sites. It is important that data
sources and decision methodology be carefully documented in making
this selection. Because these sites will likely be carefully scrutinized
in the final selection process by others (especially landfill sites), the

investigator must have good reasons for rejecting certain sites and retaining others for further analysis. In addition, the number of potential sites selected should be kept to a minimum. A greater number of candidate sites results in more computation effort in the systems analysis.

1. Landfill Sites

In order to accurately compare the life-cycle economics of a resource recovery facility which will operate for twenty years or more with the landfilling option, a twenty-year landfilling system must be established. If the landfills currently utilized by the study area are approaching capacity limits, expansion of existing sites or the location of new sites will be necessary within the study period. If new sites are required, the potential locations of new landfill capacity must be selected so that waste transportation costs can be established for the landfilling system.

The final selection of a new landfill site is technically complex and politically sensitive. A complete discussion of the proper methods for landfill site selection would require much more space than is available for this discussion. Comprehensive siting methods have been developed which utilize citizen participation and rating systems (52,53). However, for a relatively simple selection of an initial set of alternative sites for further analysis, the following steps should be taken.

a. Calculate Acreage Requirement - The number of acres needed for the new (or expanded) landfill site is dependent on a number of factors including the capacity remaining in existing fills, the waste generation rate (tons per year), the in-place compaction factor (pounds/CY), the waste-to-cover ratio, the total fill depth, and other site-specific design requirements. Certain references (54,55) discuss these types of calculations in detail. As an example, however, a community of 100,000 people generating waste at 5 pounds/capita/day would require about 136 acre-ft. per year of landfill space assuming an in-place density of 1000 pounds/cubic yard and a waste-to-cover ratio of 4:1. Further, assuming a 15-year capacity, an average fill depth of

30 feet, and an allowance of 10 percent of the total land area for roads, buildings, trench side slopes and perimeter zones, a total of about 75 acres of land would be required. The investigator should consult the references and local or state regulatory agencies to obtain the correct factors for use in the study area in question.

b. Identify General Site Areas - In general, expansion of an existing landfill site is preferable to locating a new site. If such an expansion is not possible because of space limitations or other reasons, sites with the following characteristics should be selected:

1. Vacant industrial or agricultural land
2. Relatively close to the study area
3. Favorable soil type and topography
4. Good access to major roads
5. Surrounded by industrial or agricultural land uses--maximum separation from residential areas
6. Separation from surface and ground water
7. Not in a flood zone
8. Not within 10,000 feet of an airport with jet traffic

By compiling three different types of maps for the study area and surrounding land, general site areas can be selected which meet these criteria. Existing and future land use maps are usually available in comprehensive plans for cities or regional areas. Maps showing topography, most buildings, and major roads are available from the U.S. Geological Survey (USGS). The 7 1/2 minute quadrangle maps are the most suitable. Soil survey maps (for the top five feet of soil mantle) for a major portion of the country are available from the Soil Conservation Service of the U.S. Department of Agriculture. State geological and soil service agencies, and university departments of soil sciences and geology should also be consulted for information.

By simultaneously examining these three map types, the investigator should first look for vacant land in agricultural or industrial land use zones as close as possible to the study area. Present and future land

use maps should be examined to eliminate those areas near airports
with jet traffic (because of the potential for attracting flocks of birds),
or those scheduled for residential or commercial development. Those
areas which are more than a mile from a major highway along an
unimproved road should also be seriously considered for elimination
because of the high cost of access improvement. The soil type in each
remaining area should be determined next. Detailed discussions of the
most suitable soil types should be consulted (53). In general, fine-
grained soils such as inorganic clays of low to medium plasticity,
gravelly clays, sandy clays, silty clays, and lean clays will provide a
good combination of workability and low permeability. Sites with
coarse-grained soils or highly organic soils should (with some exceptions)
be avoided. Sites with low elevations which include streams, ponds,
marshes, waterways, or other surface water should be avoided (most
states have specific regulations on separation distances from bodies of
water).

   c. Ex..ine Specifc Parcels - County Assessor's records of
landown:rs within each general area selected from the maps in the
previous step should be consulted next. The investigator should, if
possible, find a parcel of sufficient acreage with a single owner
(multiple ownership creates more complexity in acquisition). The parcel
should be at least 1000 feet from any existing residence. Local real
estate brokers should also be consulted to identify any parcels currently
for sale. By limiting the number of selected sites to between two and
five, computation effort in the systems analysis and in future detailed
site selection activities can be minimized.

   2. Transfer Station Sites

The objective of a transfer station is to provide a convenient drop
point for collection vehicles when a distant landfill or resource recovery
facility is utilized. As such, potential locations should be close to the
centers of waste generation activity. Although not as politically
sensitive as the siting of landfills, the final site selection for a transfer

station can be complex. The following are guidelines for selecting a set of potential sites for further analysis.

A site adjacent to an existing landfill which is to be closed in the future is usually the most preferable site if it is close enough to the center of waste generation in the study area. However, it is not advisable to locate the transfer station on closed landfill cells because of foundation problems. By utilizing a location near an existing landfill, incompatibility with surrounding land use is avoided since a properly-designed transfer station operation will generate fewer noise, odor and visual problems than a landfill. In addition, waste haulers are familiar with the transportation routes and persons living and working along routes, used by the collection vehicles will not experience a change in traffic volumes.

In addition to existing landfill sites, the investigator should look for vacant industrial land with good access to major transportation routes. Existing land use and zoning maps from city or county records should be examined to locate these sites.

Generally, two to five acres is required for a transfer station depending on its capacity. USGS quadrangle maps can be examined for the topography on each selected site. A sloping site is preferable, especially if there is access from both a low and high elevation (such as a corner lot). The County Assessor will have records of the ownership of each parcel of land.

A total of two to five potential sites should be sufficient for further analysis. If the study area is long and narrow in shape, it may be wise to select sites at both "ends" for possible multi-site configurations.

3. Resource Recovery Facility Sites

The siting of certain types of resource recovery facilities depends on the location of the selected energy buyer. Facilities which generate steam (or co-generate steam and electric power) must be located as close as possible to the buyer of the steam. Selection of the sites for these technical alternatives, therefore, must await the results of the energy market analysis (see Chapter 3). Many of the site criteria given in the following paragraphs, however, also apply to those facilities constrained to the vicinity of the energy buyer.

Facilities which generate electric power or which produce a fuel (solid, liquid, or gas) for sale to a distant facility are not directly constrained to the location of the energy buyer. It is necessary, therefore, to select a set of alternative locations which can be analyzed from a raw waste transportation cost standpoint in the systems analysis. Many of the considerations previously discussed for transfer station sites also apply to resource recovery sites.

The site of an existing transfer station, or adjacent to an existing landfill is generally the most preferable for the reasons stated previously. In addition, the investigator should examine land use and zoning maps to locate vacant industrial land with good transportation access relatively close to the center of waste generation activity. Ten to twenty acres is required depending on the type and capacity of the facility. Availability of utilities (gas, water, electric power, sewer) and adequate fire protection should also be selection criteria. The availability of cooling water (from a well, river, or lake) is critical to the location of an electric power facility. Certain areas may have restrictive air quality regulations which can be a major obstacle for an electric power generation facility. Topographical information for potential sites should be obtained from USGS maps.

III. RIVER CITY EXAMPLE

The study area for the River City analysis is not strictly defined as the incorporated area as shown previously in Figure 1.2 in Chapter 1. In recognition of the growth projected to occur in the metropolitan area over the next several years, it was decided to include certain areas to the west, southwest and east in order to account for this population growth. The study area boundaries will be given in a following section.

A. Waste Generation

1. Existing System

The current situation in River City was explained in Chapter 1. Table 2.11 shows the collectors and disposal sites currently utilized. All residential waste is collected by a private hauler under contract to the city. Waste from about 60 percent of the contractor's collection routes goes to the transfer station for transport to the city's Central Landfill. The remainder is transported directly to the county-owned North Landfill. All commercial and industrial waste is privately collected and hauled to either the North or South landfills. The special wastes are privately collected and hauled mainly to the North Landfill (except for hazardous wastes which are transported out of state to approved disposal sites).

2. Existing Quantities

Scale records are available at North Landfill and at the Transfer Station. Since the Central Landfill accepts only waste coming from the Transfer Station, quantities going to this landfill are defined. Weight data for the South Landfill were estimated by the private operator from load counts and compaction factors. Existing weight data for each landfill and waste type are given in Table 2.12. Note that because many commercial and industrial loads are mixed together in one collection vehicle, it was not possible to obtain separate commercial and industrial waste data.

3. Existing Composition

The existing composition displayed in Table 2.13 was derived without a landfill sampling survey because of study budget limitations. Landfill sampling surveys conducted in other cities in the same regional area

TABLE 2.11

River City Existing System

Waste Type	Collector	Disposal Site(s)[c]
Residential	Contractor [a]	Central [d], North
Commercial	Private [b]	North, South
Industrial	Private [b]	North, South
Special/Bulky	Private [b], City	North

[a] Collection service performed by a private hauler under contract to the city.
[b] Individual citizen or business performs collection using internal labor, or by contract to a private hauler.
[c] See Figure 1.2, for locations.
[d] All waste coming to the Central landfill is transshipped through the transfer station.

along with EPA data were used to establish waste composition. It is recognized that a waste sampling program may be required in the future before final design of any resource recovery facility which may be selected.

4. Unit Waste Factors

Separate unit waste factors for residential and commercial/industrial waste were calculated because of the availability of waste data in these categories and the availability of both population and employment projections by zone. Using the residential waste given in Table 2.12

TABLE 2.12
River City Existing Solid Waste Quantities
(tons/yr)

Landfill	Total Waste	Residential	Commercial/Industrial	Special/Bulky	Total Processable[a]
North	230,300	51,700	146,400	32,200	198,100
South	72,000	60,000	12,000	0	72,000
Central	81,400	81,400	0	0	81,400
Totals	383,700	193,100	158,400	32,200	351,500

[a] Residential plus Commercial/Industrial

TABLE 2.13

River City Existing Solid Waste Composition

Component	Percent by Weight
Paper	45.4
Misc. Organics (plastic, wood, rags, yard waste, etc.)	35.8
Total Combustible[a]	81.2
Glass	8.1
Ferrous Metal	8.2
Aluminum	0.7
Other Metals	0.3
Misc. Inorganic	1.5
Total	100.0

[a] Sum of "paper" and "misc. organics."

along with an existing population of 588,000, a residential unit waste factor of 1.8 pounds/capita/day was derived. The commercial/industrial unit waste factor is 3.3 pounds/employee/day based on the commercial/industrial waste total for the study area given on Table 2.12 and total existing employment for the study area of 265,200.

5. Unit Waste Factor Projection

Scale records for the past four years for residential waste were correlated with total population estimates for the same years. The unit waste factors increased slightly during this period (1 percent to 2 percent per year). In order to ensure a conservative estimate of future waste quantity, however, the present unit waste factors are assumed to remain constant over the study period.

6. Future Quantities

A projection period of 20 years (1985-2005) was selected assuming that the first full year of operation of a resource recovery plant would be 1985 if implementation planning was commenced in 1980. Population, total employment, and transportation time and distance data were available in five-year increment projections for the transportation zones used by the regional planning agency with jurisdiction in River City. The 290 zones which cover the study area were grouped into 42 Waste Generation Districts (WGDs). Figure 2.2 shows the boundaries of the WGDs which define the study area as compared to the incorporated area shown in Figure 1.2, Chapter 1. Note that presently unincorporated areas are included in the study area to account for future growth.

Table 2.14 displays the processable solid waste projections based on the unit waste factors previously derived and the population and employment projections summed for all WGDs in the study area. Note also that Table 2.14 gives a projection of the Special/Bulky waste category (non-processable) for use in landfill sizing based on a constant percentage of the total processable number in each projection year.

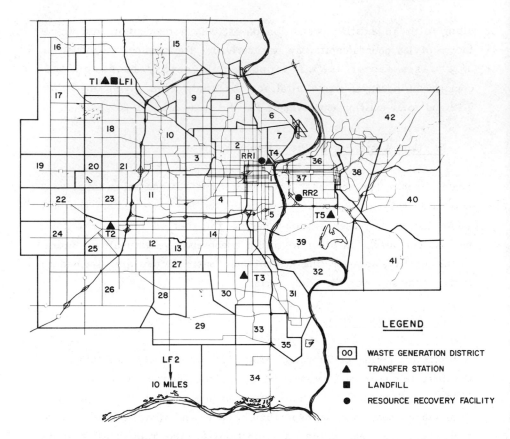

Figure 2.2. River City study area with WGD boundaries and potential facility locations.

7. Future Composition

Composition of the waste stream is assumed to remain essentially unchanged during the study period. No pending mandatory deposit legislation exists, and expansion of existing source separation programs will not significantly affect composition in the future.

B. Haul Costs

Unit haul costs for collection vehicles (primary haul), transfer vehicles (secondary haul) and for residue/ash haul, were calculated from data

TABLE 2.14

**River City Solid Waste Quantity Projections
All WGDs**

Year	Population	Total Employment	Residential Waste (tons/yr)	Comm./Ind. Waste (tons/yr)	Total Processable Waste (tons/yr)	Special/Bulky(ton/yr)[a]
1980	588,000	265,200	193,100	158,400	351,500	32,200
1985	634,800	291,100	208,500	175,300	383,800	35,200
1990	676,200	311,350	222,300	187,500	409,800	37,500
1995	717,600	331,600	235,700	199,700	435,400	39,900
2000	759,000	351,850	249,300	211,900	461,200	42,200
2005	800,400	372,100	262,900	224,100	487,000	44,600

[a] Based on 9.16% of Total Processable from 1980 data on Table 2.12.

made available by the city's contractor, and from equipment manufacturers. Tables 2.15 through 2.17 give the derivations of the dollars/mile and dollars/hour unit haul costs for each vehicle. Note that separate unit haul costs are established for residential and commercial/industrial waste collection vehicles. The commercial/industrial unit haul costs will be applied only to the commercial/industrial waste quantities generated in each zone to account for the differences between commercial route trucks, and residential route trucks.

C. Potential Facility Sites

1. Landfill Sites

The cumulative solid waste quantity (processable and non-processable) for the period 1985 to 2005 is about 9 million tons (see Table 2.14). Assuming certain standard factors for landfill design, about 350 acres would be needed (including allowances for roads perimeter buffer zones, etc.) for a 20-year landfill.

Existing landfills were examined first for possible expansion. Only the county-owned North Landfill is suitable for expansion because it is surrounded by agricultural land and is more than a mile from any residential development. No surface water exists in the surrounding area, and experience in the existing landfill operation has revealed excellent soil types for a landfill operation. Expansion of the Central Landfill is not possible because it is an older operation which has become surrounded by commercial and industrial land uses. The private operator of the South Landfill is under heavy pressure from local residents to close the operation because residential development has concentrated in this southwest portion of River City. Expansion of the South Landfill is, therefore not politically feasible.

An additional site was also selected for analysis. A new major private landfill is currently under development south of the study area about 10 miles. This new site is large enough to accommodate all of River City's waste, and has been granted a state solid waste permit.

TABLE 2.15

River City
Collection Vehicle Transport Cost Derivation
(1980 $)

Item	Cost
1. Residential Waste: 25 CY, Side Load Packer, 6-ton payload	
Cost Per Mile:	
Depreciation[a]	$ 0.34
Maintenance and Tires[b]	$ 0.35
Fuel[c]	$ 0.20
Insurance[d]	$ 0.20
Total $ per mile	$ 1.09
Labor Cost Per Hour (3-man crew)[e]	$26.45
2. Commercial/Ind. Waste: 30 CY Front Load Packer, 8-ton payload	
Cost Per Mile:	
Depreciation[f]	$ 0.39
Maintenance and Tires[b]	$ 0.35
Fuel[c]	$ 0.20
Insurance[d]	$ 0.20
Total $ per mile	$ 1.14
Labor Cost Per Hour (1-man crew)	$ 9.25

[a] Capital Cost for truck and packer body = $57,000, assume 50% is for truck only. Assume 5-year life, 15,000 miles per year, 10% salvage value.
[b] Avg. maintenance cost = $0.30/mile, tire cost = $170 per tire x 10 tires per vehicle plus 10 recaps at $80 each = $2,500 for 50,000 miles of service.
[c] Avg. 5 miles per gallon during transport made, $1.00 per gallon (diesel fuel).
[d] Avg. of $3,000 per vehicle per year, 15,000 miles per year.
[e] Driver wage = $19,200/year, Laborer Wage = $17,860/year (including benefits, overhead and management). Use 1 driver and 2 laborers for a 3-man crew, and 1 driver for a 1-man crew, 2,080 hrs. per year.
[f] Capital Cost for truck and packer body = $65,000, assume 50% is for truck only. Assume 5-year life, 15,000 miles per year, 10% salvage value.

TABLE 2.16

River City
Transfer Vehicle[a] Transport Cost Derivation
(1980 $)

Item	Cost
Cost Per Mile:	
Depreciation[b]	$ 0.24
Maintenance and Tires[c]	$ 0.20
Fuel[d]	$ 0.25
Insurance[e]	$ 0.10
Total $ per mile	$ 0.79
Labor Cost Per Hour[f]	-

[a]   75 CY, compacted trailer with semi tractor.
[b]   Capital Cost for tractor and trailer = $79,000, expected life of 325,000 miles, no salvage value.
[c]   Maintenance cost = $0.12 per mile, tire cost = $170 per tire x 18 tires plus $80 per recap x 18 = $4,500 for 60,000 miles.
[d]   Average of 4 miles per gallon, $1.00 gallon (diesel fuel).
[e] $4,500 per vehicle per year, 45,000 miles per year.
[f]   Driver labor is included in the fixed annual cost of the transfer station rather than in these unit haul costs.

Operation is scheduled to begin within one year.

Figure 2.2 shows the locations of the two selected landfill sites (LF1 and LF2).

2.   Transfer Station Sites

Five potential transfer station sites were selected for further analysis. Figure 2.2 gives their locations. T1, T3, and T4 are at the North Landfill, South Landfill, and the existing transfer station sites respectively.

TABLE 2.17

River City
Residue/Ash Vehicle[a] Transport Cost Derivation
(1980 $)

Item	Cost
Cost Per Mile:	
Depreciation[b]	$ 0.53
Maintenance and Tires[c]	$ 0.32
Fuel[d]	$ 0.20
Insurance[e]	$ 0.13
Total $ per mile	$ 1.18
Labor Cost Per Hour[f]	$ 9.23

[a] 16 CY Dump Truck.
[b] Capital Cost = $40,000, expected life of 75,000 miles, no salvage value.
[c] Maintenance cost = $0.30 per mile, tire cost = $100 per tire x 6 tires per vehicle plus 6 recaps at $50 each = $900 for 50,000 miles of service.
[d] Average of 6 miles per gallon, $1.20 per gallon (gasoline).
[e] $2,000 per vehicle per year, 15,000 miles per year.
[f] One driver at $19,200 per year (including benefits overhead and management), 2,080 hrs. per year.

T2 was selected because it appeared that a transfer station would be needed in the southwestern portion of the city where significant residential and commercial development has been occurring. The location is a parcel of vacant land in an existing industrial park with excellent rail and interstate highway access.

T5 was selected because it appeared that a transfer station might be needed on the eastern side of the river. Because there are only three river crossings in River City, transportation times and distances between the two sides of the river are unusually large. In addition, the city has already been considering the location of a transfer station in the industrial area chosen for the site.

3. Resource Recovery Facility Sites

Two potential sites for resource recovery facilities which are not constrained to energy market locations were selected. They are shown as RR1 and RR2 on Figure 2.2.

RR1 is at the site of the existing transfer station. It was chosen because it is currently being utilized as a solid waste facility and is located in an industrial area near the downtown central business district and a major power generation plant. Also, the existing transfer station building and tipping area can be expanded into a resource recovery facility which produces electric power or a fuel for burning in the nearby power plant. Another advantage is that most of the land required is already owned by the city.

RR2 is a parcel of vacant city-owned land next to an existing sewage treatment plant which has been closed. The site is on the eastern side of the river in a newly developed industrial park near another major electric power plant. Further chapters in this book will continue the River City analysis based on these basic data.

REFERENCES

1. D. Rimberg, Municipal Solid Waste Management, Noyes Data Corp., Park Ridge, N.J., 1975.

2. U.S. Environmental Protection Agency, Omaha-Council Bluffs Solid Waste Management Plan-Status Report 1969, SW-3tsg, U.S. Government Printing Office, Stock No. 5502-0012, Washington, D.C., 1971.

3. Henningson, Durham & Richardson, Inc., Northwest Dallas Receiving and Processing Facility for Solid Waste-Pre-design Report, 1975, prepared for the City of Dallas, Dept. of Streets and Sanitation, May 1, 1975.

4. Henningson, Durham & Richardson, Inc., unpublished analysis of scale records from the Fairfax County, Virginia, I-66 Landfill, 1979.

5. Henningson, Durham & Richardson, Inc., unpublished analysis of scale records from the landfills operated by the City of Phoenix, Arizona, 1976 through 1979.

6. J.P. Woodyard and A. Klee, Solid Waste Characterization for Resource Recovery Design, Proc. Sixth Mineral Waste Utilization Symposium, U.S. Bureau of Mines, and IIT Research Institute, May 2-3, 1978.

7. N. Chin and P. Franconeri, Composition and Heating Value of Municipal Solid Waste in the Spring Creek Area of New York City, Proc. 1980 National Waste Processing Conf., ASME, May 11-14, 1980, p. 239.

8. H.I. Hollander et al., A Comprehensive Municipal Refuse Characterization Program, Proc. 1980 National Waste Processing Conf., ASME, May 11-14, 1980, p. 221.

9. U.S. Environmental Protection Agency, OSW, Fourth Report to Congress-Resource Recovery and Waste Reduction, SW-600, Washington, D.C. 1977.

10. U.S. Environmental Protection Agency, St. Louis Refuse Processing Plant: Equipment, Facility, and Environmental Evaluations, EPA-650/2-75-044, May, 1975.

11. U.S. Environmental Protection Agency, Evaluation of the Ames Solid Waste Recovery System, Part 1: Summary of Environmental Emissions; Equipment, Facilities, and Economic Evaluations, EPA-MERL, grant number R803-903010, Oct., 1977.

12. D.J. Hagerty, J.L. Pavoni, and J.E. Heer, Jr., Solid Waste Management, Van Nostrand Reinholt, New York, 1973.

13. J.L. Pavoni, et al., Handbook of Solid Waste Disposal, Van Nostrand Reinholt, New York 1975.

14. W.L. Bider and W.E. Franklin, A method for Determining Processible Waste for a Resource Recovery Facility, Proc. 1980 National Waste Processing Conf., ASME, May 11-14, 1980, p. 211.

15. Henningson, Durham & Richardson, Inc., County of Fairfax, Solid Waste: Energy Resource Recovery Study, April, 1979.

16. Henningson, Durham & Richardson, Inc., Solid Waste Energy and Resource Recovery for Pinellas County, Florida, May 1976.

17. Henningson, Durham & Richardson, Inc., City of Charlotte, N.C., Solid Waste Disposal and Resource Recovery Study, Vol. 1, Nov. 1978.

18. Lake County, Ill. Dept. of Public Works, Engineering Report on Solid Waste Disposal, Vols. 1, 2, and 3, Sept. 1971, and May 1972.

19. Wayne County Dept. of Public Health, Wayne County Solid Waste Management Plan, 1976.

20. Henningson, Durham & Richardson, Inc., Downriver Solid Waste Energy and Materials Recovery Feasibility Study, Sept. 1976.

21. Henningson, Durham & Richardson, Inc., Assoc. of Grand Rapids Area Governments and West Michigan Regional Planning Commission, Resource Recovery Study, Final Report, June, 1977.

22. Henningson, Durham & Richardson, Inc., State of Montana, Solid Waste Management Strategy, 1976.

23. Henningson, Durham & Richardson, Inc., Dubuque County, Iowa Solid Waste Resource Recovery Study, Dec., 1975.

24. Henningson, Durham & Richardson, Inc., City of Phoenix, Arizona, Solid Waste Energy and Materials Recovery Study, 1978.

25. Henningson, Durham & Richardson, Inc., Strategy for Solid Waste Management, North Central Texas Region, North Central Texas Council of Governments, May, 1974.

26. Henningson, Durham & Richardson, Inc., Metropolitan Area Planning Agency, Resource Recovery Feasbility Study, Draft, April, 1980.

27. T.V. DeGeare and J.E. Ongerth, Empirical Analysis of Commercial Solid Waste Generation, Journal of the Sanitary Engineering Division, Proc. ASCE, December, 1971, p. 843.

28. Metcalf and Eddy of Michigan, Inc., Southeast Michigan Council of Governments Solid Wastes Study, Detailed Report, March 30, 1973.

29. Vilican, Lehman, and Assoc., Roscommon County, Michigan Solid Waste Management Plan, 1975.

30. Henningson, Durham & Richardson, Inc.,/Harry S. Johnson Companies, Solid Waste Energy and Resource Recovery Study for Hennepin County, Minn., Final Report, Dec. 1975.

31. Capitol Consultants, Michigan State Solid Waste Management Plan, 1972.

32. Henningson, Durham & Richardson, Inc., Racine County, Wisconsin, Solid Waste Management Plan, Final Report, 1976.

33. Henningson, Durham and Richardson, Inc., The Des Moines Story, prepared for the City of Des Moines, Iowa, 1968.

34. Henningson, Durham & Richardson, Inc., Solid Waste Management Plan for the Bi-State Metropolitan Area, 1972.

35. Henningson, Durham & Richardson, Inc., Phoenix Solid Waste Resource Recovery Implementation - Phase I, Final Report, Jan. 1980.

36. Henningson, Durham & Richardson, Inc., Solid Waste Energy Recovery and Materials Reclamation Feasibility Study for DeKalb County, Georgia, Dec. 1975.

37. F.A. Smith, Quantity and Composition of Post Consumer Solid Waste: Material Flow Estimates for 1973 and Baseline Future Projections, Waste Age, April, 1976, pp. 2-10.

38. U.S. Environmental Protection Agency, Baseline Forecasts of Resource Recovery 1972 to 1990, MRI Project No. 3736-D, Kansas City, Missouri, March, 1975.

39. Midwest Research Institute, Market Analysis for Recovered Materials and Energy from Solid Waste in Oakland County, Michigan, Final Report, Jan. 28, 1976.

40. D. Waggoner, The Oregon Bottle Bill--What it Means to Recylcing, Compost Science Sept./Oct., 1976, pp. 10-13.

41. U.S. Department of Commerce, The Impacts of National Beverage Container Legislation, Staff Study A-01-75 (unpublished), Oct. 1, 1975.

42. U.S. Environmental Protection Agency, The Beverage Container Problem: Analysis and Recommendations, NTIS #: PB-213 341, Sept. 1972.

43. U.S. Environmental Protection Agency, Resource and Environmental Profile Analysis of Nine Beverage Container Alternatives, MRI, NTIS #: PB-253-486, 1974.

44. Federal Energy Administration, Energy and Economic Impacts of Mandatory Deposits, Research Triangle Inst., NTIS #: PB-258-638.

45. P.M. Hansen, Residential Paper Recovery - A Municipal Implementation Guide, EPA pub. no. SW-155, Washington, D.C. 1975.

46. Henningson, Durham & Richardson, Inc., unpublished analysis of compacted versus open-top transfer station designs for the Southeastern Virginia Public Service Authority, May 1977.

47. Personnal Correspondence with Mr. Al Green, Los Angeles County Sanitary District, 1977.

48. Personnal correspondence, Mr. William Shaw, City of Harvey, Illinois, 1977.

49. Personal correspondence, Mr. William Hall, City of Seattle, Washington, 1977.

50. Personal correspondence, Mr. R. Basagbnishon, City of Bel-Aire, California, 1977.

51. Personal correspondence, Mr. George Tattrie, City of Wyandotte, Michigan, 1981.

52. Henningson, Durham & Richardson, Inc., Solid Waste Disposal Facilities for the City of Dubuque, Iowa - Preliminary Engineering Report, 1974.

53. Henningson, Durham & Richardson, Inc., City of Charlotte, N.C., Solid Waste Disposal and Resource Recovery Study - Vol. II, Potential Facility Locations, 1978.

54. U.S. Environmental Protection Agency, Sanitary Landfill Design and Operation, Brunner and Keller, SW-287, NTIS #: PB-227-565.

55. S. Weiss, Landfill Disposal of Solid Waste, Noyes Data, Park Ridge, N.J., 1975.

CHAPTER 3

MARKETS

I. INTRODUCTION

In an analysis of the economic feasibility of resource recovery, an examination of the markets for recovered energy and materials is necessary for two major reasons. First, the available markets determine which forms of energy and which categories of materials should be recovered. This determination, in turn, defines the alternative recovery technologies which should be considered. There is little wisdom in building an expensive facility which recovers materials or energy for which there is no long-term, viable market. Therefore, it is essential to determine which recovered resources are the most valuable to potential buyers, what specifications are required by the buyers, and what value buyers place on these resources. Second, the available markets help determine the location of the resource recovery facility. In some cases, the location of the energy market will dictate the location of the facility (steam markets). Therefore, when resource recovery is being analyzed, both the locations of waste generating areas and the locations of the buyers of recovered resources (especially energy) must be known.

Although the correct apporach to a resource recovery feasibility study is to find the markets first, then examine technologies which are defined by the markets, the investigator should have a basic knowledge of the available technology before beginning the market survey. A preliminary screening of technologies may assist in limiting the scope of the market survey. For example, the investigator may wish to eliminate from consideration the "mass burn" technology because of air quality restrictions in the study area; or the investigator may wish to eliminate composting technology because of its history of difficulty in obtaining long-term purchase commitments for compost, or pyrolysis because of remaining questions regarding its technical viability. The market survey and selection of alternative resource recovery technologies are, in actual practice, performed simultaneously. In an actual feasiblity study there is an on-going give-and-take between the results of the market survey and the analysis of available technology. These issues are, however, dealt with separately in this book (Markets-- Chapter 3, and Alternatives--Chapter 4) in order to maintain orgainzational clarity. For the reasons stated above, it is recommended that the investigator read both Chapters 3 and 4 before beginning the market survey.

The following sections describe how to calculate the quantities of recoverable resources using the basic data gathered in Chapter 2 and how to select a set of potential energy and materials markets. The last section continues the River City example for numerical illustration.

II. QUANTITIES OF RECOVERABLE RESOURCES

Recoverable resources in processable solid waste can be divided into two general categories: materials which can be separated and burned or chemically converted to recover energy (the combustibles), and those materials which can be separated for use as a soil conditioner (compost) or as secondary materials to supplement virgin materials in the manufacture of new products (ferrous metals, glass, aluminum, plastic, paper). Some materials such as paper and plastics could be classified in both categories.

The quantities of materials which are potentially recoverable from the available waste stream must be calculated in advance of the market survey.   This calculation is especially important for the energy market survey.

A. Energy

The amount of energy available in various forms from processable solid waste is dependent upon both the characteristics of the waste and the specific energy conversion technology utilized.   With the waste composition defined as described in Chapter 2 along with a general knowledge of available technology, estimation factors with a reasonable degree of accuracy can be defined in the absence of knowledge of the exact size and type of energy conversion equipment to be utilized. Table 3.1 displays certain typical energy conversion factors which may be utilized at this stage of the investigation.   The investigator should confirm that the assumptions used in Table 3.1 are applicable for the study area in question before utilizing these factors.

B.   Materials

The amounts of various materials recoverable from the waste stream are a function of the waste composition and the recovery efficiency of the separation technology.   Using the U.S. EPA national waste composition (4) and typical separation efficiencies, Table 3.2 displays material recovery factors which can be used with a reasonable degree of accuracy.   The investigator should utilize local composition data when making these calculations, if available.

Note that plastic and paper recovery are not included in Table 3.2. Although experimental separation technology exists for these materials, there are no full-scale plants which separate these materials in a marketable form from mixed municipal refuse in operation at present.

TABLE 3.1

Energy Conversion Factors

Energy Form	Rough Quantity Yield Per Ton of Waste	Rough Energy Yield Per Ton of Waste (Btux10^6)
Solid Fuel[a]	.75 to .84 tons	8.1 to 8.6
Pyrolysis Gas[b]	17,741 scf	7.9
Pyrolysis Oil[c]	42.1 gallons	4.8
Steam[d]	6,650 lb.	6.7
Electric Power[e]	667 KWH	--

[a] Shredded, screened, air-classified (1)
[b] Pyrolysis temperature @ 900 degrees C, 447 Btu/scf (2)
[c] Garrett system, 113,910 Btu/gallon (3)
[d] 4,500 Btu/lb. raw waste, 50% to 70% combustion efficiency, 10% in-plant usage, 1,000 Btu/lb. steam.
[e] 4,500 Btu/lb. raw waste, 15,000 Btu/KWH total plant heat rate.

TABLE 3.2

Material Recovery Factors

Material	Yield Per Ton of Solid Waste (lbs.)	Reference
Compost[a]	750	(5)
Ferrous Metals[b]	155	(6)
Aluminum[c]	11	(7)(8)
Glass[d]	104	(9)

[a] 25% of raw waste is non-compostable, 2:1 reduction ratio in the digestion process.

[b] 8.6% of raw waste, 90% recovery efficiency

[c] 0.7% of raw waste, 80% recovery efficiency based on data collected at Recovery 1, New Orleans, using an Eddy Current Separator and vibrating screens.

[d] 10.4% of raw waste, 50% recovery efficiency for non-color-sorted glass using froth-flotation equipment.

Note also that materials which can be source separated (newsprint, ferrous, and aluminum cans, office paper, etc.) are not included in Table 3.2 since this book's scope is limited to post-disposal resource recovery.

III. IDENTIFICATION OF POTENTIAL MARKETS

The next step in the market analysis, after the quantities of energy and materials available from the waste generated in the study area are estimated, is to select a set of potential markets which express interest in buying these recovered resources, and which meet certain technical criteria as explained in the following sections.   In this step, the investigator will also determine initial market values for the various energy forms and recovered materials.

A. Energy

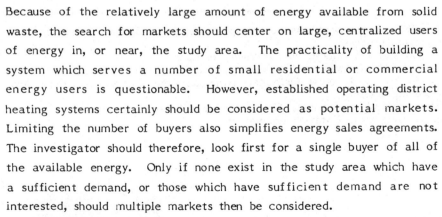

Because of the relatively large amount of energy available from solid waste, the search for markets should center on large, centralized users of energy in, or near, the study area.   The practicality of building a system which serves a number of small residential or commercial energy users is questionable.   However, established operating district heating systems certainly should be considered as potential markets. Limiting the number of buyers also simplifies energy sales agreements. The investigator should therefore, look first for a single buyer of all of the available energy.   Only if none exist in the study area which have a sufficient demand, or those which have sufficient demand are not interested, should multiple markets then be considered.

There is considerable interest in certain large U.S. cities in the economic development potential of resource recovery facilities.   For example, the Port Authority of New York and New Jersey has developed a master plan for the construction of several resource recovery industrial parks for the purpose of attracting industrial development into economically depressed areas of New York and New Jersey (10). While this concept has considerable merit in certain cities, it approaches the question of the feasibility of resource recovery from a perspective different from the methodology presented in this book. The perspective of this book is to analyze the feasibility of resource recovery as an element in the solid waste management system for a certain study area. As such, the methodology presented calls for the identification of markets in advance so that the facility can be designed to meet the specific needs of <u>existing</u> industrial or utility buyers.

1. Market Survey

Large users of energy rely in most cases on the use of electric power purchased from utilities, the direct firing of fuel in the manufacturing process (steel rolling mills, cement manufacturers, etc.), or the production of steam for process use or power generation by burning various fuels. The market survey should not concentrate on finding larger electric power users because the expense in building a solid waste fired electric power plant with the on-line reliability required by large industrial electric power user makes the cost of electric power generally higher than utility-supplied power. For this reason, sale of electric power is usually limited to utilities which can use other boilers on its power grid to obtain required levels of reliability. Direct-fire applications for the various forms of fuels recovered from solid waste in various industrial processes are still experimental. Therefore, the market analysis should not concentrate on this type of buyer.

The market survey should focus on locating local electric power utilities and large users of fuel for steam generation. This steam is used by utilities to generate electric power, and by industries as either process steam (steam used as an energy source in a manufacturing process), or for in-plant space heating and cooling. By locating large steam boilers, therefore, information on both local utilities and industries can be gathered. Steam may also be used in large institutions such as hospitals or schools for laundry, dishwashing, heating, and cooling. Markets exist either for fuels derived from solid waste which can replace the conventional fuels used to produce the steam, for steam at specified temperatures and pressures to supplement or replace that which is now produced, or for electric power for sale into the utility grid.

Much of the information required for an energy market survey can be obtained by investigating the existing boilers in or near the study area. These boilers use various feedstocks depending on the application. Some can fire only gas or fuel oil, some can fire coal or other solid fuels in addition to gas or fuel oil, and yet others are old coal-fired units which have been converted to fire gas or fuel oil. The purpose of the market survey is to obtain the characteristics of all of the large boilers within a reasonable distance of the study area. These data can be the basis of selecting the best market for all recovered energy forms.

The most comprehensive source of information about boilers in a given area is usually the list of all permitted point sources of air pollution usually kept by the state agency responsible for air quality. In addition to giving the location, rated capacity and owner, many times these lists are often computerized and updated annually so the types of fuels and the amounts of each used in each year are also available. Reference 11 gives a more comprehensive discussion of the data available and how it can be used. It is impractical, however, to contact every owner of a boiler on these lists because a large number of small boilers are usually included. The investigator should first eliminate from consideration those establishments with a combined heat input to all boilers of less than about 25 million Btu per hour. This is roughly equivalent to the energy in 50 tons per day of processable waste. For larger study areas (over 100,000 population) this limit should be raised to about 100 million Btu per hour in order to limit survey time.

The location of those establishments with boilers above the capacity limit should be examined next. Only locations either within the study area or within 10 to 20 miles should be retained for further analysis. If the air quality permit list is not available, the state or local chamber of commerce will have a list of the largest industries in the study area. The headquarters of the local electric utility can be consulted for the locations and technical data for all area power plants.

An initial telephone contact can be helpful in further eliminating certain potential markets from the remaining list. For electric utilities, the investigator should talk to someone in facilities engineering who can answer questions about how many hours a particular power plant is operated each year, what fuels are used, and what boiler designs exist at the site. If a utility power plant under consideration for a recovered energy user is only scheduled to operate a few hours out of the year during peak periods for instance, this prospect should be removed from the list. When telephoning a prospective industrial or intitutional energy buyer, the investigator should ask to speak to the plant engineer or the director of the physical plant. The telephone call should be used to confirm the capacity information which may have been available on the air quality listing, and also to discuss briefly the fuels used by the boilers, steam conditions, and seasonal steam demand quantities.

Personal visits are recommended for all potential markets remaining under consideration following the telephone interviews. During these visits the investigator can examine the boiler equipment and steam piping layout, and can identify any available land space on site for resource recovery facilities. Specific information about each boiler, and historical data on steam usage should also be gathered during these visits. A questionnaire form such as the example given in Appendix A can be filled out by the investigator during the plant visit or left with the plant contact person to fill out. The information gathered should include plant location; types and amounts of fuels used on an annual, quarterly, or monthly basis if possible; current fuel prices; boiler type, rating, age, ash handling system, steam conditions, air pollution control equipment; and steam demand on a monthly basis for the past two to three years.

2. Use of Survey Data

The survey information gathered for each potential market is next used to identify potential markets for each energy form.

a. Markets for Solid Fuel - Solid fuel derived from solid waste is most generally the light combustible fraction cut into small particles. This fuel is many times called RDF (refuse derived fuel) and in some applications RDF has been "pelletized" or "densified." Proprietary technology also produces a dry, fine powder RDF (see Chapter 4 for a more complete discussion of solid fuel technology). Only boilers with solid fuel firing capability can utilize this kind of fuel. Potential energy buyers with coal-fired or converted coal units (which can be converted back to coal economically) should be considered for potential markets. Solid fuel derived from solid waste has been successfully co-fired in spreader-stoker type units at a nominal rate of 50 percent of the input energy in both shredded and pelletized form (12,13). Solid fuel in shredded and powdered form has been successfully co-fired with coal in pulverized coal, suspension-fired boilers at rates of up to 10 percent of input energy (11). These nominal co-firing rates can be used by the investigator in calculating the potential amount of RDF which may be utilized by each market.

b. Markets for Pyrolysis Gas or Oil - Pyrolysis gas and oil are the products of two different types of pyrolysis technology. The pyrolysis oil is formed in a process which utilizes a lower temperature than the gas process (see Chapter 4 for a more complete discussion of pyrolysis technology). Pyrolysis gas and oil can directly replace natural gas and No. 6 fuel oil, but the Btu replacement ratios are not 1:1. Pyrolysis gas has 50 percent to 65 percent less heating value per cubic foot as natural gas. This means that between two and three times as much pyrolysis gas is required to meet the same demand now served by natural gas (depending on the pyrolysis process used). Pyrolysis oil has about 75 percent of the heating value per gallon as No. 6 fuel oil. Certain modifications to natural gas burners are required to obtain the correct air-to-fuel ratio when converted to pyrolysis gas. It should be noted that questions still remain regarding the long-term technical viability of pyrolysis technologies.

Any industry or utility which presently burns either natural gas or fuel oil is a potential market for pyrolysis gas or oil. Because pyrolysis gas is not generally mixed with natural gas, a separate pipeline distribution system is required. A single industrial or utility user which can utilize all of the gas produced is therefore preferred. The investigator should utilize the appropriate replacement ratio when making these calculations. Similarly, since pyrolysis oil requires a distribution system (by truck, rail, or pipeline), a single market is preferred. The investigator should also seek a market which presently burns primarily No. 6 fuel oil since pyrolysis oil chemically approximates this petroleum-based fuel. Conversions of No. 2 fuel oil burners are, however, possible for burning pyrolysis oil.

c. Markets For Steam - As an alternative to purchasing replacement fuels, many large energy users prefer purchasing steam. The purchase of steam relieves the uncertainty of burning solid waste derived fuels in boilers designed for other fuels. Steam can be generated in a variety of different ways. Basically, the two approaches are "mass-burning" in which raw waste is burned with minimal prior processing and various systems which process the waste before combustion in specially-designed boilers (see Chapter 4 for a complete discussion of steam generation technology).

Because of the high cost of the steam distribution system required, a single market with an annual and seasonal demand sufficient to purchase all of the steam available from the solid waste is preferred. The energy conversion factor for steam given in Table 3.1 can be used for rough calculations of available steam; however, more specific calculations utilizing the buyer's required steam conditions, a specific technology, and the buyer's seasonal demand variation should be performed before estimates of annual steam sales are made for revenue calculations.

The investigator should consider industrial steam users, institutional steam users, and electric power utilities as possible markets. Since boiler manufacturers have suggested an upper limit of 750 degrees F for steam generation in waste-fired boilers (14) to avoid corrosion problems, many utilities are reluctant to utilize waste-fired steam in existing turbines designed for higher temperatures. The utility market is, therefore, limited. However, some utilities have expressed interest in modifying existing turbine equipment so that solid waste-fired steam can be utilized (15).

d. Markets for Electric Power - Electric power can be produced from solid waste with addition of turbine generators to the steam generation facilities described in the previous section. Generally, higher steam temperatures and pressures are required for generating electric power (see Chapter 4).

As previously described, the most viable approach is to sell electric power to the public utility, investor-owned utility, or municipal electric system in the study area rather than directly to industrial users. The passage of the Public Utility Regulatory Policies Act (PURPA) of 1978 and the adoption of the associated Federal Energy Regulatory Commission (FERC) regulations in February 1980 has made these markets more accessible to resource recovery facilities (16). The FERC rules:

1. require electric utilities to purchase electric power from qualifying co-generation facilities and small power producers (including solid waste and biomass fueled plants) at a rate reflecting the cost that the purchasing utility can avoid as a result of obtaining energy and capacity from these sources, rather than generating an equivalent amount of energy itself or purchasing the energy or capacity from other suppliers;

2. requires electric utilities to furnish power to these qualifying plants at non-discriminatory rates; and

3. exempts certain qualifying cogeneration and small power production facilities from certain existing regulations.

The responsibility for interpretation and implementation of these FERC rules lies with each State regulatory agency and nonregulated utility. The Public Utilities Commission of the State of California has ruled (17) that an "Avoided Cost" formula must be used for determining the rates for payment by a utility for co-generated, biomass, or solid waste power. Estimates of these rates in California for 1980 range from 37 to 45 mills/KWH depending upon time of delivery (17).

e. **Markets for Co-generation** - Co-generation is the simultaneous generation of electric power (or mechanical energy) and useful thermal energy (steam or hot gases). Chapter 4 includes a more complete discussion of co-generation technology. Solid waste co-generation systems typcially involve the production of high pressure superheated steam which is used to turn a turbine generator. Lower pressure steam is either extracted from the turbine or used directly from the turbine exhaust for industrial process steam.

The potential markets for the steam and electric power which can be produced in a solid waste-fired co-generation facility are the same as previously discussed for these energy forms separately. Co-generation should only be considered, however, if the only market for steam has a demand smaller than the steam which can be generated from the available solid waste, or the best steam market has a highly seasonal steam demand which periodically drops below the seasonal steam availability. This is because in most cases, when compared on an equivalent basis, the revenue from steam will be higher than from electric power (assuming the supply of "firm" steam to an industrial user).

3. **Selection of the Best Market**

Although not necessary, the selection of one "best" market for each energy form is preferable because the complexity of the economic analysis is simplified by doing so. In any case, the list of potential markets should be screened as much as possible before expending the considerable amount of effort required to properly analyze each market. The selection of markets from the list of available potential buyers is dependent upon three factors: 1) market demand for energy, 2) the location of the market, and 3) the value of energy to the buyer.

Market demand is important because it is generally preferable to select the market with a demand high enough to purchase all of the available energy in all seasons. If there is more than one market which meets this criterion, other factors should be used to make the selection.

The location of the market is important from a waste and/or fuel transport cost standpoint. Generally, a market closer to the study area is preferred over one a longer distance away. In the case of the sale of electric power, of course, this factor is not as critical if a number of locations for interconnections to the electric power grid exist.

The importance of selecting the market which will pay the highest price for the recovered energy is obvious. However, the investigator should discuss reliability requirements and other issues which may affect the cost of the resource recovery facility before selecting the market offering the highest energy value. If the market requires extremely high reliability, unusual delivery conditions, or guarantees of operating efficiencies, the extra costs of building and operating the resource recovery facility may offset the higher energy value.

B.   Materials

1. Separation Technology and Market Availability

The marketability of recovered materials in many cases depends on the technology used to separate the materials from the waste stream. Separation technology is well-developed for some materials, while the technology for separation of other materials has yet to be developed so

that the product can be marketed.   The following discussion is limited
to mechanical separation technologies for mixed processable waste.   A
more complete discussion is included in Chapter 4.

Separation technology of mixed municipal waste for compost is well
developed and microbial composting techniques after separation have
proved technically feasible.   The absence of a viable compost market is
the major reason for the failure of past facilities (4,18).   Although new
areas for market development have been suggested (19), composting can
not be viewed at present as a major component of a solid waste
management system mainly because of marketing problems.

The technology for magnetic separation of ferrous metals from the
solid waste stream is well developed (see Chapter 4) and stable markets
are available for these metals.   The recovery of aluminum and other
non-ferrous metals has also been demonstrated, but these processes are
not without problems.   Difficulties with equipment reliability and
product purity are still being encountered, though these are gradually
being resolved.   (See Chapter 4 for a more complete discussion.)   There
is a strong market for recovered aluminum which meets certain purity
requirements.

Several methods for glass recovery are under development.   Froth
flotation techniques for example, have been demonstrated which can
produce a 99 percent+ pure product which meets specifications for
mixed-color cullet.   Mixed color cullet can be used in the manufacture
of glass containers, but not to the extent that color-sorted cullet can
be used.   An optical color sorting device is currently under
development; however its ability to separate materials that meet
manufacturer's specifications has not yet been demonstrated.   (See
Chapter 4.)

Plastics make up only a small percentage of the waste stream, and
must be separated by parent resin type to be recyclable.   The
technology for this separation from mixed municipal waste has not yet
been developed.   Recovery techniques are expensive and established
markets for rubber and plastics are not available (20,21).   Mechanical
separation of a marketable paper fraction from mixed wastes has been

attempted, but has not proved feasible. For the above reasons, plastics and paper markets are not discussed in this chapter even though markets for source-separated paper exist.

2. Market Requirements

a. Ferrous Metals - Basically there are two types of markets for the ferrous metals (largely steel cans) recovered from solid wastes: chemical processing and remelt applications. The chemical processing markets include tin recovery and copper precipitation. Remelt markets include steel mills, foundries and ferro-alloy producers. When selling steel scrap, it should be remembered that buyers' prices assume material is delivered to the buyer, and prices are usually quoted on a gross ton (2,240 pounds) basis, rather than a net ton (2,000 pounds) basis. Each potential market is discussed in the following sections. Included is a description of the process and the applicable material specifications.

i. The steel remelt market - The use of steel cans as a scrap material is not new, but until recently only very small percentages were used for each melt. When national interest in resource recovery increased, the steel industry conducted controlled melting tests to ascertain the most efficient levels for remelting scrap cans in their furnaces. At the present time, scrap steel cans, which compose about 90 percent of the ferrous metals found in municipal residential wastes, can be melted in electric furnaces, basic oxygen furnaces, open hearth furnaces, and even blast furnaces.

Probably more external factors affect the market price of steel scrap than any other commodity. Some of the many which are difficult to control by the steel scrap seller are: short delivery cycles, availability of gondolas, discriminatory freight rates, and supply of metallurgical coke.

The yield is the amount of steel actually recovered during melt and is proportional to the amount of ferrous materials in the recovered product. Steel companies will base their pricing for the product on an anticipated yield. Failure to meet the prescribed percentage will reduce the income to the seller. Thus, if a 50-ton load of "ferrous" products is shipped by the seller to a steel furnace and the actual yield is only 80 percent, the seller will be penalized the 20 percent by direct loss in revenue. In addition, the seller will have incurred the penalty of the freight charges paid on 10 tons of non-ferrous materials included in the shipment. The steel company will be penalized because of increasing charge time, production losses, slagging problems, and possible air pollution problems. The seller must consistently deliver the best possible ferrous products to the steel remelt buyer.

The presence of tin, carbon and copper is the most critical element of concern. Generally, aluminum slags in processing, thereby separating from the ferrous material, but other elements are alloyed. The acceptable levels vary between processes, plants, and types of steel product made. Generally, refuse derived steel, without detinning, can be used to make most new carbon steel products. Although every type of steel-making furnace can accommodate steel recovered from solid wastes (without detinning), electric furnaces probably are the most common users.

A high bulk density is mandatory. Many companies insist on bales or bundles and size varies with the individual plant. Some prefer a "nuggetized" or briquetted product. Generally, a bulk density of 75 pounds per cubic foot is required for remelt.

ii. The tin recovery (de-tinning) market. De-tinning is an industrial process which recovers tin from tin-plated steel. The usual sources are cans rejected in the can manufacturing process, cans recovered from municipal solid waste before incineration, and others. There are about 7.5 pounds of tin in each ton of scrap cans. The de-tinning industry is this country's only domestic source of tin. De-tinners indicate they will buy all the clean, nonincinerated steel can scrap they can get. Basically, tin recovery is an alkaline chemical process in which scrap is treated with a hot caustic soda in the presence of an oxidizing agent, such as sodium nitrate, and the tin is dissolved as sodium stannate. The tin is then recovered from the solution as sodium stannate crystals, as metal by electrolysis, or as a tin oxide.

A high tin content is desirable, and aluminum from bimetal beverage cans can be accommodated only in certain plants. Thus, the presence of heavy metals and aluminum tends to degrade the value of ferrous metals for this market. The product must not be tightly baled or compressed. De-tinning depends on chemical reaction on the surface of the metal, so a high surface-to-volume ratio is desirable. An ideal product consists of uniformly sized "coupons" of constant thickness.

iii. The gray iron foundry market. Iron foundries represent another potential market for ferrous metals recovered from municipal refuse. Metal contaminants, normally found in refuse-derived steel, can generally be accommodated by foundries making gray (rather than ductile) iron castings. These contaminants are primarily tin, aluminum and lead. The tin level is sufficiently low that it is not a problem in the production of gray iron castings. Aluminum and lead also usually exist in acceptable concentrations (22).

The product should be shredded and not baled. Although a dense product is desirable, most foundry cupolas cannot accommodate baled material. This factor should be evaluated with each prospective buyer of the material to verify specific requirements for baling or shredding, since final processing cost is a consideration.

iv. The ferro-alloy industry market. A fourth market for refuse derived steel is in the production of ferro-alloys, where iron is combined with carefully controlled amounts of other elements, such as silicon and manganese. The resultant materials are then used as an additive in melts for alloy steel, as castings in foundries, and other industrial applications.

One principal ferro-alloy providing a market for ferrous scrap is ferrosilicon, in which the iron source is ferrous scrap. Traditionally, either machine shop turnings or incinerator scrap has been used, depending upon the economic factors of supply and demand. However, for metallurgical reasons, incinerator scrap or unincinerated refuse-derived steel normally should not exceed 50 percent of the input material.

Since elemental iron is the crucial product, yield is very important. There are apparently no problems with the quantities of copper, tin, aluminum, etc., normally found in refuse-derived steel. The product must be shredded so that it can be used in existing material handling equipment. This requirement precludes baling. The density of such a product should be as high as possible to minimize long distance rail shipping costs.

v.   Copper precipitation iron market.   Reclaimed cans with all impurities removed have a ready market in the copper industry as "precipitation iron." The process involves the use of sulphuric acid in a chemical reaction which leaches out the copper from low-grade ores in the form of copper sulphate. This process is one of the few economically feasible ways of recovering copper from low grade ores. Existing copper precipitation plants are located in the southwestern part of the United States.

The product must be free of tin, lead and aluminum. Copper presence is not objectionable. The ratio of oxides to ferrous metal is also critical. The product must not be baled or tightly compressed. The precipitation process depends on the chemical reaction on the surface of the metal, so high surface to volume ratio is desirable. An ideal product consists of uniformly thin "coupons."

vi.   Secondary materials dealers.   Scrap metal dealers and brokers in the study area may also represent a large potential market for metals recovered from solid wastes. While this sector of industry is not a final consuming industry, the scrap metal dealers provide products to other consuming industries, as well as to export buyers. The importance of this industry to the potential resource recovery facility operator could be significant because scrap metal dealers can provide an extremely useful function as brokers for these materials. In this role, the dealers could provide ready and continuing markets for metals, process them as required, and resell to any of the ultimate buyers which offered the best current pricing structure. In other situations, the scrap metal dealer could provide additional processing, such as shredding, nuggetizing, baling, de-tinning and burning which might be necessary to meet the alternative buyers' specifications. Also, the scrap metal dealers could provide transportation of the recovered metals, whether by truck, rail or ship.

Secondary materials dealers will purchase scrap metals of many types, provided that they are generally clean and representative of a good metal product. In some cases, this type of buyer may prefer to broker the material directly to the ultimate users or may process and further beneficiate and accumulate the material before delivering it to the ultimate buyer. In many instances, the secondary materials specifications will be determined by the ultimate buyer and would therefore determine the specifications required by the secondary materials dealer.

b. Nonferrous Metals - Although nonferrous metals comprise a small portion of municipal solid waste (seldom exceeding one percent by weight), once segregated these metals are extremely valuable. If these products are segregated and in a pure form, there are many alternative buyers for them. However, buyer interest depends to a large extent upon the purity of the product and the segregation method used. Since current technology permits the isolation of an "aluminum only" product, alternative buyers for this product have been explored, as well as those for the mixed nonferrous metal product.

i. The aluminum market. Aluminum can be separated from solid wastes after the wastes have been shredded and air classified, and ferrous metals have been magnetically removed. The glass and aggregate material should also be removed by screening. Aluminum has also been successfully recovered from incinerator ash in certain tests (see Chapter 4).

Recovered aluminum is currently being purchased by two market types: major aluminum producing companies, and aluminum smelters producing specification ingots. Beverage cans separated manually and collected by volunteer recycling groups (as well as aluminum can stock recovered from mixed refuse) are remelted by the major aluminum companies to make new beverage can stock. Similar remelt markets exist for industrial aluminum scrap which is segregated by alloy.

Mixed alloy aluminum scrap is purchased by aluminum smelters. These dealers melt the scrap and adjust the metallurgy as necessary to produce billets of lower alloy aluminum. There are many such smelters in this country and abroad, purchasing industrial scrap and aluminum recovered from car shredding operations. These buyers can accept the lower purity aluminum recovered from solid wastes by the heavy media process. These are also buyers of the high purity material sought by the major aluminum companies.

In addition to these basic markets for recovered aluminum, there are local brokers (scrap dealers) who purchase such material and ship it to whichever basic market offers the highest price at that time.

Aluminum recovered from waste will meet various industry grades, depending upon the method of separation. A top-quality, refuse-derived aluminum product should be at least 95 percent aluminum. Nonmetallics should be minimal because of their effect on yield. The composition of the product should be essentially aluminum cans. A high incidence of other aluminum products will result in down-grading of the material.

Depending upon the buyer finally selected, the product may be shipped loose or in low-density bales in railcars or trucks (the latter is generally the rule). Material should be dry and covered during shipment. Certain buyers will accept the material F.O.B. receiving facility, so this point should be clarified before calculating revenues.

ii. Mixed nonferrous metal market. Mixed nonferrous metals can be recovered from solid wastes as a byproduct, or an extension, aluminum recovery. This mixed-metals fraction (generally less than 0.25 percent of the wastes) consists of minute quantities of copper, lead, zinc, stainless steel, coins, keys, residual aluminum, and numerous other metal products that are neither magnetic nor aluminum. Each constituent is valuable individually, but extensive further processing to isolate the individual metals is generally not feasible economically.

Probably the best outlet for this material is a local scrap dealer, particularly one operating an auto shredder, and recovering nonferrous as well as ferrous metals. These companies, in turn, may hand-pick some of the higher valued materials such as copper and sell the rest or incorporate the product with a load of nonferrous scrap ultimately scheduled for the export market.

The market is neither strong nor well-defined, but a mixed nonferrous metal product is of interest to the secondary materials industry. The types and quantities of metals present in the product will govern the market value. Analytical assays must be performed before pricing formulas can be established.

c. Glass Markets - Recovered glass can be used as cullet material, which is remelted for the manufacture of new glass containers. This concept has been demonstrated by virtually every glass container manufacturer using glass bottles and jars collected manually and color-sorted by volunteer recycling centers. Such color-sorted, used glass is in demand periodically by the glass companies.

In practical terms, the major technical difficulty for any operator of a resource recovery facility lies in implementing a system that can meet the purity and color-sorting requirements and accommodate the costs of moving the product to market. An acceptably pure glass fraction can be obtained by mechanical means and/or froth flotation techniques. However, the only mechanized color-sorting technique available is an optical comparator which is undergoing development testing. The results (23) have shown that a color-sorted fraction which meets glass container manufacturers' specifications for refractory contamination levels has not yet been acheived.

There are, however, markets for noncolor-sorted, refuse-derived glass. When a glass company has a furnace which produces green or amber glass, some amount of noncolor-sorted glass cullet can be used safely in the mixture. This material is added to the batch up to the "allowable impurity" level, which may be as high as 20 percent of the mix. This material can be derived from refuse using the froth flotation process, which evolved from the mining industry.

The Glass Container Manufacturers' Institute specification for noncolor-sorted glass has the following features. The product must be dry (no drainage), noncaking, and free flowing with not over 0.25 percent organics, not over 0.05 percent magnetic material, and maximum size of 1/4 inch. The glass mixture itself must be totally soda-lime (container) glass. One hundred percent must pass a 2-inch by 2-inch mesh screen, and 10 percent maximum may pass a 140 mesh screen. Solid inorganics, other than metal or refractory and nonmagnetic metals in the 20 mesh range shall not exceed one particle in 40 pounds of mix, maximum size 1/4 inch. Basically, the above mixed cullet material can only be used to manufacture amber or green glass.

3. Market Survey Methods

The investigator should begin the materials market survey by compiling a list of local as well as national potential markets. Local scrap dealers can be identified from the telephone yellow pages. Local industrial buyers can be identified from lists provided by the local chamber of commerce and other sources. The investigator should look for aluminum producers and smelters, detinners, glass container manufacturers, steel making facilities, and other nonferrous metal smelters. The U.S. EPA has compiled a national list of materials buyers (24) which can be consulted for potential markets outside the study area; however, this list cannot be considered comprehensive.

After developing this list, letter inquiries should be sent to potential buyers. The letter should provide a brief explanation of the market analysis being performed and the approximate material quantities and specifications available. The material recovery factors given in Table 3.2 can be used for rough quantity calculations. The letter should request an indication of the buyer's interest in purchasing the recovered materials, the chemical and physical specifications which

TABLE 3.3

River City – Energy Recovery Potential

Energy Form	Units	1985	1995	2005
Solid Fuel[a]	Tons/yr	287,900	326,600	365,300
	Equivalent tons/yr of coal	174,600	147,000	164,400
Steam[b]	lb./hour continuous	291,400	330,500	369,700
Pyrolysis Gas[c]	10^6 scf/yr.	6,800	7,700	8,600
	Equivalent Nat. Gas (10^6 scf/yr)	3,000	3,500	3,900
Pyrolysis Oil[d]	10^6 gal/yr	16.2	18.3	20.5
	Equivalent No. 6 Fuel Oil (10^6 gal/yr)	12.4	14.0	15.7

Energy Form	Units	1985	1995	2005
Electric Power[e]	10^6 KWH/yr	256	290	325
	Average Load (KW)	29,200	33,200	37,100

a 75% of raw waste @ 5400 Btu/lb., coal @ 12,000 Btu/lb.
b 6,650 lb steam per ton of raw waste, continuous.
c 17,741 scf per ton of raw waste @ 447 Btu/scf, Nat. gas @ 1000 Btu/scf.
d 42.1 gallons per ton of raw waste @ 113,910 Btu/gallon, No. 6 oil @ 148,840 Btu/gallon.
e 667 KWH per ton of raw waste.

they require, and a general indication of the value of the materials. Follow-up telephone calls may also be necessary to secure information.

Another approach which has been used (25) is more extensive. The approach involves a limited local survey and the negotiation of advance commitments to purchase materials via binding Letters of Intent (LoI). This approach is more applicable in larger communities in which local primary buyers are likely to exist (heavy industrial buyers). This approach is also more applicable to those projects in which it is imperative to have firm materials purchase agreements in advance of system design. The investigator is encouraged to consult reference 25 for more detail regarding this method.

IV.   RIVER CITY EXAMPLE

A.   Quantities of Recoverable Resources

Using the composition data given in Table 2.13 in Chapter 2 and the total projected quantites of processable waste given in Table 2.14 of Chapter 2, quantities of recoverable resources were calculated for River City.   Tables 3.3 and 3.4 display the calculated data for energy and materials quantities projected in ten-year increments from 1985 to 2005.   The factors given previously in Tables 3.1 and 3.2 were used in these calculations.   Other assumptions are listed in the table notes. Note that in Table 3.3, the equivalent quantities of conventional fuels are tabulated where appropriate for comparison.

B.   Potential Markets

1. Energy

The State Air Quality Board was contacted for a listing of data on all permitted point sources of air pollution.   Utilizing this list, and a

TABLE 3.4

**River City - Materials Recovery Potential
(tons per year)**

Energy Form	1985	1995	2005
Light Fraction[a]	287,900	326,600	365,300
Compost[b]	143,900	163,300	182,600
Ferrous Metals[c]	28,300	32,100	35,900
Aluminum[d]	2,100	2,400	2,700
Glass[e]	15,500	17,600	19,700

a 75% of raw waste, mostly combustibles separated by screening, shredding, and air density separation.
b 37.5% of raw waste.
c 8.2% of raw waste, 90% recovery efficiency.
d 0.7% of raw waste, 80% recovery efficiency.
e 8.1% of raw waste, 50% recovery efficiency, mixed-color cullet.

knowledge of the available energy quantities as given in Table 3.3, the list of potential markets was reduced to four. Personal visits and discussion were held with each. Table 3.5 lists the data gathered for each potential market, and Figure 3.1 shows market locations.

a. River City Power & Light (RCPL), North Station - The RCPL North Station is a 500 MW power plant along the river in the northeastern section of the study area built between 1955 and 1960 (see Figure 3.1). Coal is the primary fuel at this facility, however, natural gas is fired periodically. No oil firing capability exists. The plant is base-loaded, so the seasonal demand variation is minimal. Steam is generated at 900 psig, 900 degrees F in the existing units.

Because of the existing high steam temperature, the sale of steam from a refuse-fired boiler is not possible because of boiler corrosion problems (see Chapter 4), but this establishment is a potential market for RDF. Electric power sales to the RCPL grid is also a potential market. At a supplementary rate (on an input Btu basis) of 10 percent RDF at this facility, all of the RDF generated in the year 2005 could be burned. However, discussions with RCPL staff revealed that they are not interested in co-firing RDF with coal at this plant. The reasons include:

1. Retrofitting the boilers to burn RDF involves high costs in order to assure proper boiler operation

2. Increased ash disposal, maintenance requirements, and downtime attributed to RDF firing would significantly increase operating costs

3. Plans for new, larger base-loaded units to be completed within five years will change the operating schedule of the North Station from its present base-load status to "mid-load" operation with significant periods of complete shut down during the winter months.

TABLE 3.5

**River City Potential Energy Markets
Boiler Data**

Establishment	Fuels Used	Existing Equipment	Demand Variation	Comments
River City Power & Light North Station	Coal, Nat.Gas	4 boilers (coal/gas) totalling 500 MW	base-loaded	market for electric power only, not interested in RDF.
River City Power & Light East Station	Coal, Nat.Gas Nat.Gas	1 boiler, (coal/gas) 650 MW	base-loaded	market for electric power only, not interested in RDF.
River City District Heating System	Nat.Gas	3 boilers (gas/#2 oil) 125,000 pph each @ 650 psig/750 degrees F.	seasonal	market for steam, or pyrolysis gas
River City Industries	Nat.Gas	2 boilers (gas/#2 oil) 200,000 pph each @ 150 psig/sat.	Relative-ly steady	market for steam

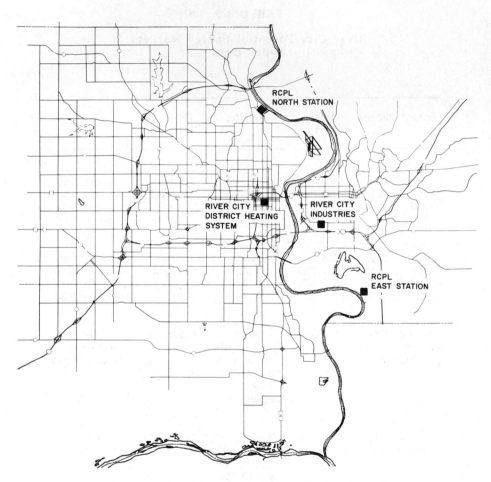

FIGURE 3.1. River City Energy Market Locations

RCPL did, however, express interest in purchasing electric power at the approximate rates as shown in Table 3.6 plus a monthly payment of $1.75 per KW of capacity.

b. River City Power & Light, East Station - The RCPL East Station is a 650 MW pulverized coal-fired plant (one unit) completed one year ago. Steam conditions are 950 psig, 1000 degrees F. The plant is located in the southeast section of the study area (see Figure 3.1). This plant is base-loaded and burns mostly coal.

Again, because of the existing steam conditions, this establishment is not a candidate steam market. RCPL is also opposed to co-firing RDF with coal in this new unit for many of the same reasons as stated previously for the North Station. As detailed previously, RCPL is interested in purchasing electric power.

c. River City District Heating System - The River City District Heating System is a privately-owned system located in the downtown central business district (see Figure 3.1). The plant supplies steam and chilled water for space heating and cooling for a number of downtown buildings. The plant has three, 125,000 lb /hr boilers (600 psig, 750 degrees F) which burn natural gas and No. 2 fuel oil. Natural gas is the primary fuel. In addition, the plant contains three steam turbine-driven water chillers rated at a total of 12,000 tons. The plant and underground pipeline system has been in place for over twenty years.

The District Heating System has no solid fuel firing capability, so it is not a potential RDF market. It is, however a market for pyrolysis gas or steam. The present gas demand is sufficient to burn all of the pyrolysis gas which could be produced from the projected year 2005 waste quantity. Natural gas is currently purchased by the District Heating System at $2.00 per million Btu. Natural gas curtailments are not expected in the forseeable future.

Peak steam demand is presently about 360,000 pounds per hour from present customers. This peak is expected to rise to about 450,000 pounds per hour with the addition of new customers before 1985. By

TABLE 3.6

River City - RCPL Electric Power
Purchase Rates 1980

RCPL Load Condition	Time Period	Annual Hours	Purchase Rate (mils/KWH)
Light Load	Ap. through Oct. 12:00 A - 7:00 A	1500	2.7
Medium Load	All other Periods	7000	9.0
Heavy Load	Mid-June through Mid-Sep.,Weekdays 3:00 P - 7:00 P	260	25.0
	Weighted Average		8.4

inspection of Table 3.3, this peak demand is higher than the year 2005 solid waste steam availability. However, during certain months, the seasonal demand is less than the seasonal steam availability. This situation occurs because periods of peak demand do not correspond with the periods of peak solid waste generation. The owners of the District Heating System expressed an interest in purchasing steam at 600 psig, 750 degrees F at a rate of about $2.50 per thousand pounds in 1980 dollars.

d. River City Industries - River City Industries (RCI) is a medium-sized manufacturer which utilizes steam as a heat source in the manufacturing process. Steam is also used by RCI for space heating and cooling via absorption chillers. As Table 3.5 indicates, RCI currently has two boilers, each rated at 200,000 pounds per hour of saturated steam at 150 psig, capable of burning both natural gas and No. 2 fuel oil. RCI is located in the East Side Industrial Park (see Figure 3.1).

Because the existing boilers at RCI do not have solid fuel firing capability, RDF cannot be utilized.   RCI is, however, an excellent candidate for a steam market.   Their current average steam demand is about 300,000 pounds per hour (annual average) which closely matches the available steam in 1985 (see Table 3.3).   RCI indicated interest in purchasing steam at a rate of about $2.50 per thousand pounds.

2. Materials

No viable markets for compost were found in or near River City. The most promising potential market appeared to be a local fertilizer manufacturer, but because of a lack of experience with the use of solid waste compost as a bulking agent, no significant interest in purchasing this material was found.

A list of potential purchasers of recovered metals and glass was compiled from data from the local telephone yellow pages, the chamber of commerce, scrap metal market trade publications, and US EPA publications.   Letter inquiries were sent to potential buyers on the list and several responses were received.   The response to the survey is summarized in Table 3.7.

In general, there appears to be a very strong market interest in recovered ferrous metals and aluminum.   For the purposes of the example, a price of $15.00 per ton F.O.B. the resource recovery plant (that is, after subtracting transportation costs) will be used for ferrous metals recovered before burning.   For ferrous metals recovered from incinerator ash (mass burn systems) a price of $5.00 per ton (F.O.B. the resource recovery plant) will be used because of higher contamination. Further, because of the purity requirements of the aluminum buyers and the fact that separation technology is only beginning to prove viable, aluminum separation will not be undertaken in the River City example. However, space for future addition of such separation equipment will be designed in to the facilities.   The interest for scrap glass is mainly for color-sorted cullet.   Since the separation technology is not yet developed for such a product, the installation of equipment to serve this market does not appear justified.

TABLE 3.7

River City - Responses to Materials Market Survey

Material	Interested Buyers	Comments	Market Value [a]
Ferrous Metal	Foundries (local and national) Detinners (national) Steel Mills (local & national) Local Scrap Dealers	Market is strong, established, diversified, and widespread	$25-$65/ton
Aluminum	Smelters (national) Local Scrap Dealers	Strong market interest, strict purity requirements	$.20-$.50/lb.
Glass	Glass Container Manufacturers (local and national) Local Scrap Dealers	Strong interest for color-sorted cullect with strict specifications on contamination	$20-$30/ton

[a] Delivered to buyer.

REFERENCES

1. U.S. Environmental Protection Agency, St. Louis Refuse Processing Plant: Equipment, Facility, and Environmental Evaluations, EPA-650/275-044, May, 1975.

2. W.S. Sanner, et al., Conversion of Municipal and Industrial Refuse into Useful Materials by Pyrolysis, U.S. Bureau of Mines, Report of Inv. No. 7428, August, 1970.

3. G.M. Mallan and C.S. Finney, New Techniques in the Pyrolysis of Solid Wastes, presented at the 73rd National Meeting AIChE, Minneapolis, Minnesota, August 27-30, 1972.

4. U.S. Environmental Protection Agency, OSW, Fourth Report to Congress--Resource Recovery and Waste Reduction, SW-600, Washington, D.C., 1977.

5. J.L. Pavoni, et al., Handbook of Solid Waste Disposal, Materials and Energy Recovery, Van Nostrand Reinholt, New York, 1975.

6. H. Alter, S.L. Natof, K.L. Woodruff, and R.D. Hagen, The Recovery of Magnetic Metals from Municipal Solid Waste, NCRR, RM77-1, Washington, D.C., 1977.

7. National Center for Resource Recovery, Inc., Aluminum Recovery from MSW Using an Eddy Current Separator, TR-80-8, Washington, D.C., June, 1980, p.12.

8. National Center for Resource Recovery, Inc., Preparation of Glass and Aluminum Concentrates Using a Double-Deck Vibrating Screen, TR-80-7, Washington, D.C., June, 1980, p. 14.

9. John Arnold, Initial Mass Balance for Froth Flotation Recovery of Glass from Municipal Solid Waste, NCRR Technical Report RR77-1, Dec. 1977.

10. S.S. Passage and G.B. Liss, Resource Recovery and Economic Development, in Proc. 1980 National Waste Processing Conf., ASME, May 11-14, 1980, p.251.

11. Jerome Kohl, et al., A Manual for Locating Large Energy Users for Co-generation and Other Energy Actions, North Carolina State University and the Research Triangle Institute, March, 1980.

12. U.S. Environmental Protection Agency, Evaluation of the Ames Solid Waste Recovery System   Part I--Summary of Environmental Emmissions: Equipment, Facilities, and Economic Evaluations, MERL, EPA-600/2-77-205, November, 1977.

13. S.H. Russell, Refuse Derived Fuel (RDF) in Proc. Engineering Foundation Conference on Present Status and Research Needs in Energy Recovery From Solid Wastes, Sept. 22, 1976.

14. D.A. Vaughan, Corrosion Mechanisms in Municipal Incinerators Versus Refuse Composition, in Proc. Engineering Foundation Conference on Present Status and Research Needs in Energy Recovery from Solid Wastes, Sept. 22, 1976.

15. Henningson, Durham & Richardson, Inc., Phoenix Solid Waste Resource Recovery Implementation--Phase I, Final Report, Jan., 1980.

16. U.S. Department of Energy, Federal Energy Regulatory Commission, Small Power Production and Cogeneration Facilities; Regulations Implementing Section 210 of the Public Utility Regulatory Policies Act of 1978, 18 CFR Part 292, Federal Register, Vol. 45, No. 38, Feb. 25, 1980, p. 12214.

17. Public Utilities Commission of the State of California, OII No. 26, Interim Order, Decision No. 91109, San Francisco, Cal., Dec. 19, 1979.

18. D. Rimberg, Municipal Solid Waste Management, Noyes Data Corp., Park Ridge, N.J., 1975, p. 70.

19. R.H. Greeley and A.R. Nollet, "New Horizons in Composting," in Proc. 1980 National Waste Processing Conference, ASME, May 11-14, 1980.

20. B. Baum and C.H. Parker, Solid Waste Disposal, Vol. 2 Reuse-Recycle and Pyrolysis, Ann Arbor Science, Ann Arbor, 1974, p. 103.

21. J.W. Jensen, J.L. Holman, and J.B. Stephenson, "Recycling and Disposal of Waste Plastics," in Recycling and Disposal of Solid Wastes, T.F. Yen ed., Ann Arbor Science, Ann Arbor, 1974, p. 219.

22. C.B. Dallenbach and R.R. Lindeke, "Utilization of Refuse Scrap in Cupola Gray Iron Production," in Proc. 5th Mineral Waste Utilization Symposium, Eugene Aleshin ed., U.S. Bureau of Mines and IIT Research Institute, April 13-14, 1976, p. 234.

23. J.P. Cummings, "Glass and Non-Ferrous Metal Recovery Subsystem at Franklin, Ohio--Final Report," in Proc. 5th Mineral Waste Utilization Symposium, Eugene Aleshin ed., U.S. Bureau of Mines, IIT Research Institute, April 13-14, 1976, p. 176.

24. U. S. Environmental Protection Agency, Market Locations for Recovered Materials, EPA/530/SW-518, August, 1976.

25. Harvey Alter and J. J. Dunn Jr., Solid Waste Conversion to Energy, Marcel Dekker, New York, 1980, pp. 111-130.

APPENDIX A

ENERGY USER QUESTIONNAIRE

PART A. GENERAL

1. Plant location (municipality) _____

2. Name and address of firm   _____

    _____

    _____

    _____

3. a. Person completing this form _____
    b. Position/Title _____
    c. Telephone No. _____
    d. Date _____

4. a. Type of firm or principal products produced _____

    b. S.I.C. No. _____
    c. Operating days per year _____
    d. Hours per day _____

PART B. ENERGY FACILITIES AND REQUIREMENTS

1. List the approximate quantities of each of the following types of
    fuel that your firm or plant uses quarterly at this location:

		1	2	3	4	Annual Total
a.	Natural gas (cf)					
b.	Fuel oil (gal)					
c.	Coal (tons)					
d.	Propane (cf)					
e.	Other (specify)					

2. Approximately what percent of the fuel listed above is used for:

    a. Heating and cooling _____
    b. Process steam _____
    c. Electricity _____
    d. Other (specify) _____

3. For each boiler located on your firm's site, please complete the following:

                                    Boiler #   Boiler #   Boiler #

a. Boiler Manufacturer

b. Boiler Rating
    (lb steam/hr or MW)

c. Steam Temperature
    and Pressure

d. Furnace Configuration
    (rectangular, divided, other)

e. Method of Firing
    (tangential, front, top, cyclone;
     if stoker, specify type)

f. Date of Initial Service

g. Plant or Unit Factor

h. Unusual Design or Operating
    Conditions

i. Fuel (pulverized coal,
    crushed coal, oil, gas)

j. Amount of Fuel Consumed at
    Rating

k. Fuel Values (Btu/lb)

l. Fuel Costs (present)

m. Bottom Ash Handling Capability
    (yes or no)

| | Boiler # | Boiler # | Boiler # |

n. Fly Ash Handling Capability
(yes or no)

o. Type and Efficiency of APC
Equipment

4. Are there any additional boilers planned for the near future? Yes ___ No ___

   a. If yes, how many _____

   b. Types and capacities _____

   c. Type (s) of fuel which will be used _____

   d. Approximate installation date(s) _____

5. Are there any chillers at the site? If yes, complete the following;

| | Chiller# | Chiller# | Chiller# |

   a. Type (steam
driven, electric,
absorption

   b. Capacity (tons
or Btuh)

CHAPTER 4

ALTERNATIVES

I. INTRODUCTION

The purpose of this chapter is to discuss methods for bringing together
the results of the analysis of the basic data and the results of the
investigation of markets for energy and materials. In this step of the
analysis, the investigation will focus on selecting technologies to match
the solid waste with the markets. The methodology presented is aimed
at selecting not only technologies, but a set of solid waste management
system alternatives for economic comparison.

Alternatives that are "system" alternatives and that are comparable
must be carefully selected. In order to perform a complete analysis of
the economics of resource recovery, the costs of all system components
(except the collection component) must be examined. It is not
sufficient to merely select alternative landfills or resource recovery
plants. The differences between the costs of raw waste transport,
landfilling costs for residues and non-processables, and revenues must
also be examined. The investigator must select alternative systems.
These alternative systems must also be comparable both with respect to
the quantity and types of waste handled, and with respect to time. It
is necessary for the investigator to be certain that all alternative
systems selected will handle all (or the same portion) of the study
area's waste (processable and non-processable) over the entire study
period. The cost of an existing landfill with five years of remaining

capacity which disposes of both processable and non-processable waste, for example, is not comparable with the cost of a 20-year resource recovery plant which only handles processable waste.

The following sections of this chapter include a brief review of the available technology for landfilling, waste transport, and recovery; guidelines for selecting system alternatives; and finally, a continuation of the River City example.

II. TECHNOLOGY REVIEW

The objective of this section is to briefly review available solid waste management technology. This review is intended to give the investigator a basic working knowledge of these technologies without discussing unnecessary engineering details. The investigator is advised to consult the reference materials as marked, for specific details about each process or system. There are many techniques for reducing the generation of waste (source reduction) and separation of waste prior to disposal (source separation). Although derserving of review, their consideration is beyond the scope of this book. Only "post-disposal" technologies are reviewed. In many cases source separation/reduction techniques can be compatible with the post-disposal technologies reviewed in the following paragraphs.

All available technology for post-disposal solid waste management can be loosely catagorized into two groups: traditional methods, and resource recovery methods (see Figure 4.1). Traditional methods are those which have been used in the past and continue to be the most widely used methods today. They are methods which place primary emphasis on disposal of solid waste rather than recognizing value in the waste stream. These methods have included open-dumping (now outlawed on a national basis), sanitary landfilling, and incineration with residue landfilling.

Resource recovery methods are those which incorporate one or more of the many resource recovery techniques which have developed

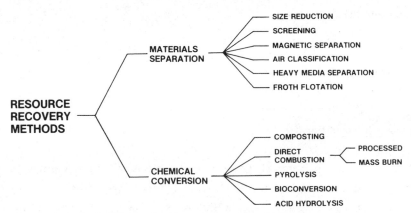

Figure 4.1.   Solid Waste Management Taxonomy

within the past decade.   Although the primary function of these methods is solid waste disposal, they sometimes place equal emphasis on recovering certain resources which are lost using traditional disposal methods.   Two general groups of technologies can be defined: materials separation, and chemical conversion.   Materials separation is the physical separation and/or sizing of the various components of solid waste, without changing the chemical makeup of any component, for the purpose of preparing them for further physical changes, chemical conversion, or disposal.   The second general group of technologies, chemical conversion, includes techniques which change the chemical composition of either processed or unprocessed solid waste for the purpose of recovering the heat released in the chemical reaction (combustion) and/or for changing the solid waste into a more useable form (such as a liquid or gaseous replacement fuel, or compost).   All of the presently available resource recovery systems are combinations of these two general groups of resource recovery technology plus certain traditional methods.   For example, a solid waste management system might include a processing plant to produce solid fuel (materials separation) for combustion in a boiler (chemical conversion) to generate steam, along with a landfill (traditional method) for disposal of residue, ash, and non-processables.

The following seven sections will discuss available technology in the areas of sanitary landfilling, volume reduction, transfer systems, materials handling/separation, composting, fuel production, and steam/electric systems.

A. Sanitary Landfilling

1. Introduction

There have been several general definitions of what constitutes a "sanitary landfill" in contrast to other, less environmentally sound methods of land application of solid wastes. In an attempt to clarify this question and to provide a basis for the identification and closure of all "open dumps," the U.S. EPA issued the "Criteria for Classification of Solid Waste Disposal Facilities and Practices" under the authority of Sections 4004 and 1008 of the Resource Conservation and Recovery Act (RCRA) of 1976, and Section 405(d) of the Clean Water Act. These Criteria, published in the Federal Register on September 13, 1979 (1) contain sanitary landfill guidelines in the following areas:

1. Elimination of open burning of solid waste

2. landfill gas control

3. safety--including control of fires, bird hazards to aircraft, and public safety

4. surface water runoff control

5. ground water contamination control

6. protection of endangered species of plants, fish, or wildlife

7. disease--including control of rodents and insects, the treatment of sewage sludge or septic tank pumpings before landfilling

8. the control of the application of solid waste to areas on or near land used for the production of food chain crops

9. the control of landfills located in floodplains

This section is concerned with the process of sanitary landfilling as defined by the U.S. EPA Criteria. The following discussion of sanitary landfilling is general in nature. Reference materials (2-5) should be consulted for more detailed discussions.

2. Landfill Methods

There are basically two methods of landfilling: the trench method, and the area method. The trench method involves the excavation of trenches which are progressively filled. Spoil from excavation is used as cover which produces an excess of cover material. With the area method, cells are constructed in all directions from a given starting point until the entire area is filled. Many layers may be placed one on top of another until the desired height has been obtained. A cell usually represents the waste received, compacted in place, and covered each day.

A combination of these two methods often works well. Trenches are dug and filled as in the trench method, after which the entire area is covered using the area method. Excess spoil from the trenches is used as cover during the area operation. The method, or methods, used in a particular application is dependent upon depth restrictions dictated by ground water levels and height restrictions dictated by aesthetics or cover material availability.

Landfill volume requirements have already been discussed in Chapter 2. Roughly 14 acre-ft. per year of fill volume is required for each 10,000 population (assuming 5 pounds/capita per day unit waste factor, 1,000 pounds/CY in-place density, and a 4:1 waste-to-cover ratio).

3. Design Considerations

A sanitary landfill can be an economically and environmentally acceptable method of solid waste disposal. Probably the greatest problem with sanitary landfilling is a lack of public acceptance. This public resistance is understandable, considering that many "sanitary landfills" have been little more than open dumps. Proper landfill design and operation can alleviate problems of odor, vermin, and litter, and may alleviate public resistance.

The two major areas of concern in the proper design of a landfill are the control of water pollution from landfill leachate and the control of the movement of decomposition gases. Measures required to prevent water pollution depend heavily upon the geology of the chosen site. If a landfill is located over a thick layer of relatively impermeable clay, channeling of surface water away from the site and using a cover material with low permeability may effectively seal the fill and prevent contamination of surface and ground waters. If existing soils are pervious, a lining of clay or plastic may be required under the fill. Control of ground water using underdrains may be required for a lined site, as well as collection and treatment of leachate generated within the fill. Lining of the site can involve large capital expenditures, so the economics should be carefully weighed before a decision is reached to develop a landfill in pervious formations.

Gases from the decomposition of wastes can create explosive or poisonous atmospheres within structures located on or near the fill such as basements and sewer lines. Methane and carbon dioxide are often formed in landfills and will migrate through the surface and sides of a fill area. Gas accumulations can be controlled by using clay barriers and by constructing gas vents in the fill. A discussion of new methods to recover and utilize landfill gas is given in a later section in this chapter.

B. Volume Reduction

Solid waste volume reduction after collection and prior to disposal has been accomplished for many years by incineration. More recently, shredding and baling have been used to reduce waste volume. These three volume reduction techniques are briefly discussed in the following paragraphs.

1. Incineration

Incineration is the controlled combustion of solid waste primarily to reduce volume. Incinerators which provide for heat recovery are in operation, but this discussion is limited to those without heat recovery. Incineration at a central plant has been practiced for more than 100 years in this country (6). Most of the incinerators which have been used in the United States were of the refractory-lined type which reduce the volume of the incoming raw waste to 15 to 25 percent of its original volume. Incinerator ash is then landfilled usually at a remote site.

As of September 1979, 64 incineration facilities were in operation in the U.S., but 101 such facilities have closed since December 1974, mostly due to inability to meet air quality standards (7). A major cause of the air pollution control difficulties is the high excess air rate which must be utilized to control combustion temperatures and to obtain complete combustion. This situation creates a high volume exit flue gas with high entrainment of particulates. The investigator should consult the reference materials for a more complete analysis of incinerator technology (8-10).

2. Shredding

Solid waste shredding as a method of volume reduction was initiated about 25 years ago in Europe where landfill space is at a premium. Some claimed advantages of shredding, due mainly to the nature of the shredded refuse, include decreased fire hazard, decreased paper blowing, and decreased vermin and bird problems. The need for cover material is decreased, particularly in instances where daily cover is not required over shredded waste. Some states require daily cover over shedded waste; others do not. Shredding and landfilling without cover (where allowed) can be particularly advantageous in areas where cover material is difficult to obtain, such as in cold climates and areas with high water tables.   Operationally, the handling of shredded waste at the landfill site is easier than unshredded solid waste.   The uniform size of the shredded waste allows it to be spread with a minimum of difficulty, to be compacted more efficiently, and it may need less cover. Shredding may, therefore, increase the life of a landfill under certain conditions.   Another advantage connected with shredding is the potential for mechanical recovery of ferrous metals from the refuse.

Disadvantages of shredding include high capital and operating costs which are only partially offset by revenues from recovered ferrous metals, and the necessity of maintaining a conventional, daily-cover landfill for items which cannot be shredded.   Fires and explosions in the shredding equipment can also increase the hazards and costs of a shredding operation.

A variety of shredding (or size reduction) equipment has been used. The most common type is the hammermill.   Shredding equipment has been used in solid waste management at over 40 U.S. locations.   The reference materials should be consulted for greater detail (4,11,12).

3.   Baling

Another volume reduction technique is the baling of waste into blocks, each weighing a ton or more.   The baling facility may be located at the landfill site or at a transfer facility where the bales are then hauled to the landfill and stacked using a fork lift.   Manufacturers of baling equipment claim that cover material is decreased, and land requirements are diminished because of decreased cover and increased density of the bales as compared to compacted-in-place raw waste. Other claimed advantages include fewer vermin problems, less litter, and better final land use, since the finished landfill has better mechanical properties which can support commercial or light industrial buildings.   In addition, less equipment is required at the landfill site. Less wear and maintenance on the landfill equipment is also claimed since it does not come into contact with the refuse.   Fuel costs may decrease because fewer pieces of equipment are needed and less movement is required on the fill.

Leachate and gas problems may be reduced in a balefill. Baled wastes do not absorb moisture as rapidly as regular landfilled wastes, and thus may not release leachate in large quantities. Furthermore, the decomposition of the waste may be retarded resulting in less methane gas produced. The landfilling of baled wastes requires specialized handling equipment, and as with shredding, the capital and operational costs are high. Since not all material can be baled, maintenance of a conventional landfill is also required. In addition to high density balers which form bales from raw waste, low density tie balers can be used to bale shredded wastes. References (4,13,14) should be consulted for more detail.

Balefill operations are located in several cities. In Omaha, Nebraska, a baling operation was initiated early in 1976. Because of the unavailability of conveniently located landfills, Omaha initiated the baling concept to reclaim a washed-out gully area for use as a park near a residential neighborhood. The operation includes a baler station located about three miles from the fill site. Included in the baling facility are two flow lines; one with a high density baler, and another line including a horizontal hammermill type shredder, magnetic separation of ferrous metals, and two low density tie balers. The shredder-tie baler line adds versatility to the plant and may be compatible with the future addition of resource recovery to the system.

The original concept included hauling bales by rail to the fill site, thereby avoiding truck traffic in the adjoining residential neighborhood. About 50 percent of Omaha's residential solid waste has been handled by the system (about 83,000 tons in 1978). System costs for 1978 totalled $17.30 per ton with a breakdown as follows: $8.82 per ton for the baling station, $4.67 per ton for rail haul, and $3.81 per ton for operations at the fill site. Initial capital costs totalled $4.5 million in 1976.

The system operated as originally intended until early 1981 when mounting citizen opposition from neighborhood residents influenced the City of Omaha to discontinue utilizing the original balefill site. Residents claimed objectionable noise, odor, and vermin as reasons for opposition. In early 1981, a private contractor was retained to transport the bales by truck to a landfill in an adjoining county about 15 miles away.

C. Transfer Stations

Transfer stations are facilities in which waste is transferred from several collection vehicles to a larger vehicle for haul to a disposal site.   The purpose is to reduce the number of vehicles hauling waste, thereby reducing cost.   The larger vehicle (usually a semi-trailer) is referred to as a transfer trailer.   A typical transfer trailer is limited to approximately 20 tons payload due to highway weight limits in most states.   This 20-ton payload may comprise loads from several packer trucks and other vehicles.   For example, three packer trucks at five tons each, plus four other trucks at one ton each, plus two pickup truck loads at 1,000 pounds each, equal 20 tons.   The material from these nine vehicles can be loaded into one transfer trailer and hauled to a disposal site more economically than if the nine vehicles were to make the trip.   Cost savings for driver labor, fuel, and the increased utilization of collection vehicle crews for collection tasks are the major areas of savings.   To be economically viable, however, it must be remembered that the costs of the transfer station added to the cost of the vehicles hauling waste from the transfer station to the disposal site must be less than or equal to the savings incurred by customers hauling to the transfer station instead of directly to the disposal site.

Some loads are taken directly to the disposal site for the following reasons:

1. The load originates closer to the disposal site than to the transfer station.

2. The individual load is a large percentage of the load carrying capacity of the transfer vehicle.

3. The load is difficult or hazardous to transfer or would cause a nuisance at the transfer station.

Most transfer stations have an area which allows collection trucks to enter and maneuver into position to discharge their loads. In some designs, trucks dump directly into a transfer vehicle in which case no storage other than extra transfer trailers is provided. In other designs, trucks discharge onto a floor or into a pit where the material is stored for a period of time. Provision must be made to reclaim material from storage if direct dumping is not practiced. Reclaiming can be accomplished with front end loaders, conveyors, overhead cranes, or other devices depending on the type of storage provided, the amount of storage, the required rate of loading and other factors. Transfer stations of all of these types are currently in operation in numerous places in the U.S.

The material from the pit or floor storage may be loaded loose into trailers or may be compacted into the trailers. To carry the same payload, a transfer trailer containing the loose material must be larger than a trailer which is compacted, but neither can legally exceed the gross vehicle load limits posted by highway authorities when traveling on public roads. Commonly, the loose trailers are in a size range of 110-114 cubic yards, whereas the compacted trailers are 65-80 cubic yards in volume.

In a floor-storage, compacted type transfer station, the waste material is pushed into the hopper of a compactor device which pushes the waste into the open rear-end of the trailer with sufficient force to compact the material in the trailer (see Figure 4.2). In the direct-loading, compaction type station, waste is unloaded directly into a compaction system. This system is usually only applicable to very small transfer stations where it is not necessary to store any waste material during peak rates of arrival. By storing material on the floor or in a pit, it is possible to reduce the number of transfer trailers and tractors by hauling the waste out at a slower rate than it is being received, but continuing the hauling over a longer period.

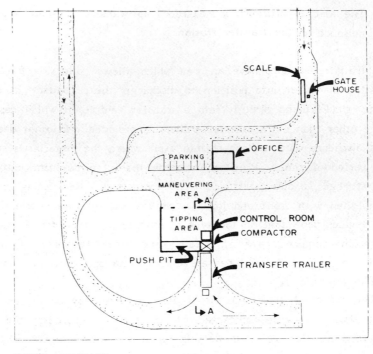

SITE LAYOUT

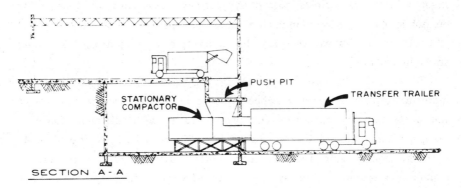

SECTION A - A

Figure 4.2.   Typcial Compacted Transfer Station (Single Compactor)

In a floor-storage, non-compacted design, waste is pushed into loading position with a rubber-tired bulldozer or front-end loader. The loading is sometimes accomplished by pushing the material to a hopper or hole in the floor with an open-top trailer positioned beneath. When the trailer is filled, it is pulled out of the loading position and a cover placed over the open top to prevent spilling enroute to its destination. Before leaving the loading position beneath the storage floor, the load is usually leveled and may be tamped down by using a relatively small hydraulic crane with a "clam shell" bucket.

In a pit-storage, non-compacted design, waste can be loaded by a crane, conveyor, or by pushing the material with a crawler tractor into a hopper with a transfer trailer positioned beneath (Figure 4.3). Again, this design would result in a non-compacted load in a trailer that may be leveled with a "clam shell" crane. The action of the heavy tractor running over the waste in the pit results in a certain measure of compaction. For this reason, the pit-storage, non-compacted design may yield higher density loads than the floor-storage, non-compacted design. The reference materials should be consulted for more detail (4,15).

D. Materials Handling/Separation

1. Introduction

A wide variety of equipment exists to move and separate processable solid waste. This equipment, which has been adapted from other material handling/separation technologies (such as mining equipment), is common to several different resource recovery technologies. The purpose of this section is to acquaint the investigator with the equipment vernacular for future reference when the various terms are used in following sections of this and other chapters of this book. This section is not intended to be a comprehensive discussion of available equipment. The investigator should consult the references as marked for details.

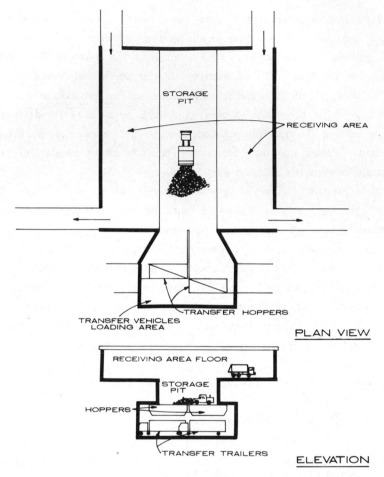

STORAGE
PIT

RECEIVING AREA

TRANSFER HOPPERS

TRANSFER VEHICLES
LOADING AREA

PLAN VIEW

RECEIVING AREA FLOOR

STORAGE
PIT

HOPPERS

TRANSFER TRAILERS

ELEVATION

Figure 4.3   Typical Non-Compacted Transfer Station (Pit Storage Type)

2. Size Reduction

A process common to many resource recovery systems is size reduction.
This is the first process in many systems, and serves the purpose of
making the waste easier to handle and separate because the maximum
particle size is controlled.   This process is basically the same as
shredding before landfilling except that the particle size produced is
smaller in many cases.   The dry process of size reduction is usually

referred to as shredding, though the terms flailing, grinding, milling, and pulverizing are also used. The most commonly-used size reduction equipment is of the hammermill type which utilizes hammers attached to a rotating shaft. The wet process of size reduction is called pulping.

Shredders are characterized as heavy equipment items with high power and maintenance requirements. Large motors are required to literally tear the refuse into pieces, and components within the machines wear rapidly from the abrasion of metals and glass. Unless large machines are used, the refuse must be pre-sorted to remove tires and appliances and other difficult-to-shred materials. Even with the larger machines, items such as cables, chain link fencing, hardened shafts and rolled carpets must be removed. Though lead times may be long, shredders are basically a "shelf" item, and are not specially designed for each application. Depending upon the materials to be shredded and the particle size required, shredders capable of handling up to 100 tons per hour are available. Detailed discussions of shredder types and performance characteristics can be found in the reference materials (8,16,17).

3. Conveyors

Conveyors commonly used in resource recovery plants include apron conveyors, vibratory pan conveyors, belt conveyors, pneumatic conveyors and bucket elevators. Steel apron conveyors are often used to handle raw refuse. They can handle impact loads and, because they can handle high tensile loads, are sometimes used in the bottom of deep storage pits. They can also be loaded from push pits or by end loaders pushing refuse off a tipping floor.

Vibratory pan conveyors can be used under shredders and as feeders for classification devices. They can level out surges from shredders for feed to air classifiers, and they can withstand projectiles from shredders which occasionally damage rubber belt conveyors.

By far the most common conveyors in resource recovery plants are belt conveyors. They can rapidly transport large quantities of materials such as shredded refuse and recovered materials. In comparison to apron and vibratory pan conveyors, belt conveyors are faster, quieter, have lower power requirements, and are less expensive. Individual components such as idlers and rubber belts can be replaced quickly and easily. They are, however, prone to damage from impact and from projectiles expelled by shredders.

Pneumatic conveyors are many times used to transport the light fraction of shredded refuse from air classifiers to storage bins and from storage bins to boilers. They are versatile in that they are not constrained to straight-line runs as other conveyors are, but they require large amounts of power, wear rapidly from abrasives (mostly glass) in the material, and are prone to plugging. Problems with plugging can be minimized by proper operation, and abrasion problems can be minimized by using easily replaceable, wear-resistant components at changes in direction (elbows).

Bucket elevators can be used to transport the heavy fraction from air classifiers. Unlike other mechanical conveyors, which are limited to about 30 degree slopes from horizontal, bucket elevators move vertically and can lift materials long distances in little horizontal space. Bucket elevators work well with granular materials, but are easily plugged by stringy and fibrous materials.

A more complete discussion of conveyor types and the performance characteristics of each can be found in the reference materials (18).

4. Screening

The simplest and most widely used means of separating different components of solid waste is on the basis of particle size. Because solid waste is a collection of materials which have different physical properties (hardness, friability, ductility), different particle sizes result from subjecting the waste stream to a size reduction operation (such as a shredder). For example, glass bottles will shatter into small pieces, while aluminum cans might only be chopped in half. Vibrating screens, rotating trommel screens, and rotating disc screens are now in use in operating resource recovery facilities (19-21). These screens are being used to eliminate small particles of non-combustible material from shredded refuse, and as components in aluminum and glass recovery systems.

Since different materials in un-shredded, raw refuse also have different particle sizes, large trommel screens to separate by particle size before shredding have also been used (21-23). The raw refuse trommel at the New Orleans, LA "Recovery I" plant has a barrel which is 46.5 feet long, with an inside diameter of 9.75 feet. The holes in the barrel are circular with a diameter of 4.75 inches. The design throughput capacity is 62.5 tons per hour, and it will accept surges of 75 tons per hour. The mass split between overflow and underflow is about 55 percent underflow and 45 percent overflow. The trommel acts to concentrate metals and glass in the underflow for more efficient recovery. The underflow also bypasses the shredder thereby reducing wear and power requirements. The trommel also acts to concentrate the combustible organics in the overflow although a significant amount of organics also report to the underflow.

5. Magnetic Separation

The removal of ferrous metals from solid waste by magnetic separation has been practiced for at least twenty years. Over thirty facilities currently utilize magnetic separation equipment for ferrous metal extraction from solid waste in the U.S. (24).

Many different configurations of magnetic separation equipment exist on the market today. Most of them are adaptations of magnets used for other recovery processes, but a few have been designed specifically for solid waste processing. Magnets may be either permanent or electro-magents. They are usually mounted in drums, conveyor pulleys, belts, or aperture type arrangements. The assembly may be suspended, in-line, crossbelt, or mounted as a head pulley in a material transfer conveyor. There are essentially two types of magentic separators available to recover ferrous metals from shredded solid waste, the belt type and the drum type. The belt type, which transports the ferrous metals under several magnets, is claimed to produce a cleaner product than drums, because the material is dropped

from one magent and then recaptured by the next, releasing nonferrous
materials which were trapped between them. Wear was formerly a
problem on the belt machines, but this was solved by armoring the belt
with metal cleats. Small head pulley magnets can be used to recover
small residual amounts of ferrous metals from the heavy fraction from
air classifiers. A complete discussion of magnetic separation technology
can be found in the reference materials (25).

6. Air Classifiers

Several manufacturers are marketing air classifiers, but all of the
machines basically feed shredded refuse at a controlled rate into a
rising column of air. The lighter, more aerodynamic particles, mostly
paper, are entrained in the rising air stream from which they are
separated in a cyclone separator. This light fraction is the fuel
fraction. The heavier particles, including rock, dirt, rubber, wood, and
metals drop out the bottom of the classifier. This material may be
landfilled or it may be separated for recovery of glass, aluminum and
other metals. A problem with air classifiers is that aluminum and
finely crushed glass often "fly" with the lighter materials. The
aluminum is lost to recovery and the glass is very abrasive to
pneumatic conveyors, storage bins and boilers.

A thorough discussion of air classifier types and performance can
be found in the reference materials (26,27).

7. Storage Systems

Both raw and processed solid wastes are extremely difficult materials
to store and reclaim. Problems with raw refuse storage include
harborage of rats and flies, decomposition odors and health hazards,
fires, inter-locking particles and low bulk density. Deep pits can store
large amounts, but reclaiming the material to uniformly feed a process
line is difficult. Ground water problems can make pits difficult and
expensive to build and maintain, and material tends to stick in corners
and decompose. Storing the material on a floor solves some of these

problems, but storage of large volumes requires large buildings. For example, floor storage of 1,000 tons may require a floor area of 20,000 square feet, including room for maneuvering of trucks and front end loaders. Raw waste storage of both the pit and floor types are currently in use in operating resource recovery facilities.

After waste is shredded, problems with inter-locking particles and low bulk density tend to increase making storage and reclaiming difficult. Shredded refuse has an angle of repose of more than 55 degrees, so side walls of storage bins usually flair out at the bottom to prevent bridging and a "live bottom" of one type or another is used. A popular storage bin for shredded refuse has a series of drag buckets which are pulled tangentially around the perimeter at the base of the pile. These buckets cut the material from under the pile and feed it into drag conveyors buried in the floor of the bin (28). Other bins use screw type devices under the pile. All types of storage/reclaim devices have high power requirements, and breakage and wear of components under the pile can present severe problems. Many of these problems are being solved as more experience is gained with operating plants.

8. Nonferrous Metal Separation

The nonferrous metal separation systems now in operation in several resource recovery plants fall into two categories: the eddy current systems, and the heavy media separation systems (HMS).

The more widely used systems are of the eddy current type. These systems operate on the principle of the induction of an eddy current in a nonferrous metalic particle by passing the particle through an electromagnetic field. This eddy current acts to move the nonferrous metalic particle away from other non-metalics in the feed stream. Some systems utilize high frequency electromagnets in various stages to effect separation, while other systems use permanent magnets in a gravity slide, or rotary drum arrangement. Size reduction, screening, air classification and ferrous metal removal is usually required to prepare the feedstock for the separator. There are several types now installed in several locations including: Ames, Iowa; New Orleans,

Louisiana; Milwaukee, Wisconsin; Baltimore County, Maryland; San Diego County, California; Bridgeport, Connecticut; and Monroe County, New York (29). These systems are, however, still experimental. Some have never worked successfully, and results to date are not conclusive regarding technical viability. The reference materials (21,30) give a more complete description of the eddy current equipment at the New Orleans facility along with its operating performance.

The HMS systems operate on the differences in specific gravity between the aluminum, glass, other nonferrous, and other non-metallic particles. Air classifier drops are usually used as a feedstock to this process. Rising current separators may also be used in advance of the HMS. The particles selectively float and sink in an aqueous slurry of ferrosilicon, magnetite, barrite, or galena. Most of the aluminum and glass particles will float while most other materials will sink (29). Additional separation by froth flotation and/or electrostatic separation is required to separate the glass particles from the aluminum product (see "Glass Recovery"). This type of system was demonstrated in the experimental Franklin, Ohio, plant for several years and has been installed in the Hempstead, New York, plant. Reference 31 has a more complete equipment description and gives operating results for the Franklin, Ohio, plant. The HMS system along with other jigging and tabling operations have also been used to separate aluminum from incinerator bottom ash by the U.S. Bureau of Mines and others (32,33).

9. Glass Recovery

Continuing efforts are being made to recover and recycle glass. Although glass is made from an abundant material (silica), other materials and relatively large amounts of energy are also required. As stated in Chapter 3, as of mid-1981, the problems of color-sorting (and others) have resulted in a situation in which the savings to a glass manufacturer from using recovered glass are not high enough to result in market prices which can support most glass separation/recycling efforts.

Equipment which has been used to separate glass from solid waste includes froth flotation, HMS, various screening systems, and electrostatic separation.   Most systems utilize a combination of these types of equipment.   Equipment to separate glass particles by color has also been developed.

The froth flotation technique capitalizes on the negatively-charged surface of glass particles.   When a small amount of a properly selected chemical commonly known as a "collector reagent" is added to a mixture of water and glass particles (three parts water to one part solids), the glass particles will become selectively coated with the collector reagent which will make their surfaces hydrophobic (water repellant).   Other non-glass particles with a lower concentration of negative surface charge are unaffected.   The coated glass particles will become attached to air bubbles when the mixture is mechanically agitated and will float to the top.   This process was developed in the mining industry and has been installed in several solid waste separation plants.   A detailed description of the froth flotation process is given in reference 34, and operating performance data for such systems can be found in other reference materials (31,35).

Heavy media separation systems were described earlier.   Froth flotation is used as an additional clean-up step behind HMS systems in some glass recovery systems.   Various jigging, crushing and screening equipment has also been used in certain applications for pre-processing of the glass fraction (36-38).

Electrostatic separators are used to separate conducting materials such as aluminum from non-conducting materials such as glass.   Glass particles will adhere to a negatively-charged drum after being positively charged while aluminum particles will fall off of the drum.   This occurs because aluminum loses its charge faster than glass.   This kind of equipment has been used as a component in the glass and aluminum separation subsystem at the Franklin, Ohio, plant (31).

The optical sorting process that has shown promise for successfully recovering a color segregated glass product from solid waste is the Sortex System.   It has been demonstrated at the EPA-sponsored Franklin, Ohio, plant operated by Black Clawson Co. and will be used at Black Clawson's Hempstead, New York, plant and one in Doncaster, United Kingdom now under construction.   The process has also been tested by the U.S. Bureau of Mines.   This process required the glass particles to be larger than 6mm (1/4 inch) in size. The first step in this process involves a separation of the opaque (mostly nonglass) particles from the transparent particles. In the second step the transparent particles are sorted by color. Photo-cells are used to measure the intensity of light transmission. The photo-cell operates a trigger to mechanically separate the glass particles. Extensive front-end processing is required prior to the color sorting process including magnetic separation of ferrous metals, screening, air classifying, and electrostatic separation.

Although this optical sorting approach is a workable system, the technology cannot yet be considered a proven technology because the color-sorted cullet from the Franklin facility failed to meet glass container industry specifications for refractory contamination. The system is reported to be improved and the ultimate test of the system will come after long-term operating results from the Hempstead plant become available. The references (8,31) should be consulted for more detail.

E. Composting

Composting is the decomposition of the organic fraction of solid waste into a humus-like substance which is used primarily as a soil conditioner. The decomposition process is accomplished with naturally occurring micro-organisms including bacteria, actinomycetes, and fungi. Size reduction and separation of non-organaics usually occurs before the solid waste is allowed to decompose. Some technologies also add sewage sludge to the solid waste before decomposition.

Various digestion techniques have been utilized including windrowing, and mechanical systems. Windrow systems place the prepared refuse in long piles, or windrows, in outdoor decomposition areas. These windrows are mechanically turned once or twice per week to facilitate aeration. Mechanical systems utilize automatic equipment to turn and aerate the prepared refuse during decomposition. Rotating drum digesters, circular tanks with rotating arms, and multi-deck digesters are some of the mechanical systems which have been used. In some systems, a finishing step is included which may comprise further size reduction, non-organics separation, and drying. Since the demand for compost is seasonal, storage space must also be allowed.

Although more than seventeen municipal solid waste composting plants have been built in the U.S., only one, in Altoona, Pennsylvania is still operating periodically. Most systems have failed not because of technological reasons, but because the compost product could not be marketed. The references (8,39,40) include more detailed discussions of composting technology.

F. Fuels Production

The following paragraphs describe resource recovery systems which produce solid, liquid, or gaseous fuels for burning in existing boilers as a primary substitute fuel, or on a supplementary basis with other fossil fuels.

1. Solid Fuels

Solid fuel systems mechanically beneficiate solid waste in order to produce a fuel which can be burned in existing boilers along with fossil fuels. Existing furnaces must have bottom ash handling equipment and particulate emissions control systems in order to burn solid refuse fuel. The term refuse derived fuel (RDF) has been widely used to identify this type of fuel. Using RDF as a supplement (the RDF only supplies part of the boiler's energy input) in existing boilers is a method of recovering energy from waste without the large capital expenditure of building a new boiler. RDF can be produced in several forms. The form produced depends heavily upon the type of boiler which will utilize the fuel. The different forms of fuel, their production methods, and their uses are explained further in the following sections. Table 4.1 gives the status of various supplemental RDF projects. Figure 4.4 exhibits a schematic diagram typical of boilers burning supplemental refuse derived fuel with coal.

a. Supplemental Fuel for Grate Firing - Many existing coal-fired boilers utilize a traveling grate inside the furnace on which crushed coal is burned. There are several different configurations of traveling grates, but all operate on the principle of allowing air to be injected through the grate from the underside to facilitate combustion. This type of coal firing mechanism is particularly adaptive to firing RDF.

In Figure 4.5, a typical RDF plant is shown which can produce supplemental fuel suitable for grate firing in a boiler with coal. There are numerous configurations of processing equipment which can be used depending on the application. Figure 4.5 is only one such configuration presented for illustration purposes. The plant has a tipping floor receiving area and primary processing with an air density separation system. In this process, the incoming solid waste is received, stored on the floor, and reclaimed from storage by pushing onto a conveyor. The shredder reduces the waste typically to a 3 to 5 inch particle size. Some processes of this type have used primary and secondary trommel screens in combination with shredders for size reduction. A magnetic separator will remove the ferrous materials which consist of the magnetic irons and steels. In this operation, the shredded solid waste, including everything except the magnetically removed ferrous is introduced to an air stream. The lightweight material will be carried in the air stream and the heavy material will drop out. The light material, consisting of approximately 80 percent of the infeed by weight, will be mostly combustible, such as paper and cardboard.

This fuel fraction is removed from the air stream in a cyclone separator and conveyed into the storage bin (or can be loaded into a transfer trailer for transport to a remote boiler). The heavy material (drops) composing approximately 20 percent of the infeed will be mostly rubber, wood, glass, dirt, rock, miscellaneous non-ferrous metals, and other material. The residue is loaded and hauled away to a sanitary

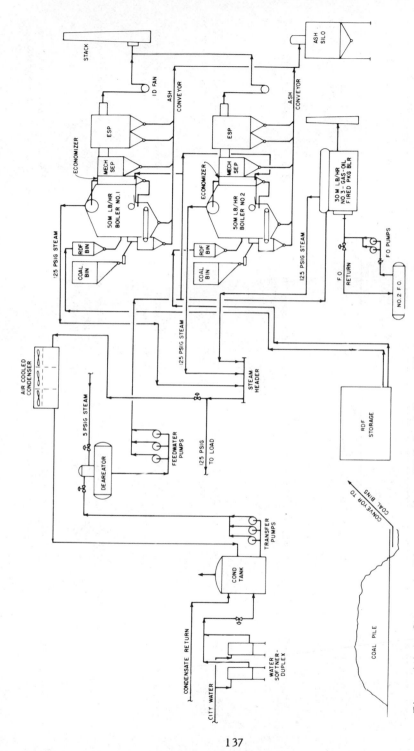

Figure 4.4. Schematic Diagram of an RDF-fired Steam Plant

137

TABLE 4.1

Facilities Producing RDF for Use as Supplemental Fuel

Location	Type of Process	Status as of 1/81[a]	Design Capacity (tons/day)	Boiler Data
Ames, IA	Shredded, air classified fuel; ferrous, aluminum recovery	Operational	200	Co-fired with coal in suspension and grate type units
Bridgeport, CT	Powdered RDF; ferrous, aluminum recovery	Started Shakedown in May, 1979, temporarily closed because vendor filed for bankrupcy	1800	Co-fired with oil in suspension
Brockton/East Bridgewater, MA	Powdered RDF; ferrous recovery	Operational intermittantly since 1977 (pilot plant)	550	Co-fired with coal, no firm, long-term market
Chicago, IL (Southwest)	Shredded, air classified fuel; ferrous recovery	Closed indefinitely, currently used as transfer station	1000	Co-fired with coal in suspension
Cockeysville, MD	Shredded air classified - fuel; fuel pellets; ferrous recovery	Operating, performed tests for EPA and Wright Patterson AFB	1200	No long-term energy Market yet
Lakeland, FL	Shredded, air classified, screened fuel, ferrous	Under construction	250	Co-fired with coal in suspension

138

Location	Type of Process	Status as of 1/81[a]	Design Capacity (tons/day)	Boiler Data
Lane County, OR	Shredded, air classified fuel; ferrous recovery	Closed indefinitely, technical problems	500	No long-term market
Madison, WS	Shredded, air classified fuel; ferrous recovery	Operational	250	Co-fired with coal
Milwaukee, WS Americology	Shredded, air classified fuel, ferrous, glass, aluminum, paper recovery	Operational at 850 tpd $4.6 million in modifications planned	1600	Co-fired with coal in suspension, slagging problems, market has stopped buying
	Shredded, air classified fuel, ferrous, glass, aluminum	In shakedown	2000	Co-fired with coal in suspension
Newark, NJ	Powdered RDF; ferrous, glass, non-ferrous recovery	Under construction, but status in question because of vendor bankrupcy	1000	

TABLE 4.1
continued

Facilities Producing RDF for Use as Supplemental Fuel

Location	Type of Process	Status as of 1/81[a]	Design Capacity (tons/day)	Boiler Data
St. Louis, MO	Shredded, air classified fuel; ferrous recovery	Demonstration plant shut down	300	Co-fired with coal in suspension
Toronto, Ont.	Shredded, air classified fuel, paper, glass, non-ferrous, compost	Experimental pilot plant operational	200	
Tacoma, WA	Shredded, air classified, fuel, ferrous	Operational intermittantly	650	No long-term market
Wilmington, DL	Shredded, air classified, fuel, ferrous, aluminum, glass, humus	Under Construction	1000	

[a] Source NCRR Bulletin, September, 1980 (46), and GRCDA Report (43).

140

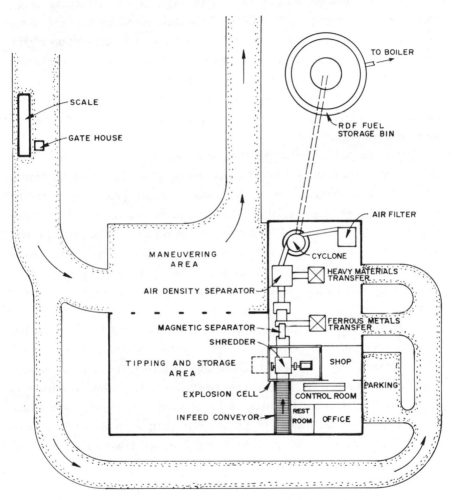

Figure 4.5.   Example Plant Layout for the Production of Grate-fired
RDF

landfill. In some processes the heavy material would be further processed to recover the aluminum, mixed non-ferrous metals and glass. Because the heavy materials removed from the fuel stream by air classification include components with high energy contents (e.g., rubber, wood, dense plastics) air classification is omitted if the extra loading of inerts does not affect boiler operation and degradation of the ash by unburned organics is acceptable. The capacity of a single-line processing plant of the type shown in Figure 4.5 could be from 20 to 100 tons per hour. The horsepower on the shredders would vary from 500 to 2000 each, depending on plant processing capacity and required particle size range.

After the supplemental fuel has been prepared, it can be marketed. If the market is remote from the processing plant, it would be loaded into transfer trailers, hauled to the boiler site and placed in a storage bin at the boiler site. If the processing plant is at the boiler site, supplemental fuel is placed in a storage bin on site. The supplemental fuel is metered from the storage bin into the boiler and blown across the traveling grate, where much of the material is burned in suspension. The remaining material burns as it moves across the furnace on the grate.

This type of solid fuel contains between 4500-5500 Btus per pound, which is approximately one-half the heat content of coal. It has a low sulfur content which makes it a desirable supplemental fuel to reduce the sulfur content in the total fuel burned. The amount of supplemental fuel usage potential is based upon the design of the boiler in which it is to be used. In certain operating systems it has been burned at a rate of 50 percent (input Btu basis) with coal for long periods without unusual boiler problems (20).

b. Supplemental Fuel for Suspension Firing - Suspension-fired boilers utilize pulverized (powdered) coal injected in an air stream into the furnace area. The small particles of coal burn as they drop to the bottom of the furnace. Some of the resultant ash is carried up with the hot gases, while the remainder becomes "bottom ash." For RDF to be burned in such a boiler, its nominal particle size must be much smaller (one inch or less) than for the grate-fired application.

Preparation of fuel for suspension firing is an extension of the grate-fired fuel preparation process. It utilizes more processing steps and may recover several materials that have potential market value. Additional processing may consist of secondary shredding, air density separation, screening and processing of materials recovered from the solid waste during air density separation. The secondary shredding is optional, depending on the size reduction in the primary shredding. Air density separation and/or screening is an essential element of the production of RDF for suspension firing. The air density separation process and screening processes have been previously described. The heavy fraction may be processed to remove aluminum, glass and other materials. Magnetic iron and steel can be removed in the primary processing and again in the processing of the heavy fraction. After materials are recovered from the heavy fraction, the residue is hauled to the sanitary landfill site.

This process is applicable where a market exists for a fuel to be burned as a supplemental fuel in an existing boiler which is outfitted to burn coal in suspension. The process may also be applicable when a market exists for a variety of secondary materials, such as magnetic iron and steel, aluminum and glass. Supplemental fuel of the type produced in this process was burned successfully in suspension at the U.S. EPA-Union Electric Company demonstration project in St. Louis, Missouri until 1975. Fuel produced by this process has been burned in suspension and on grates at the Municipal Power Plant at Ames, Iowa, on a daily basis since 1975.

The Ames processing plant (Figure 4.6) is rated at a nominal 50 tons per hour and has primary processing and secondary processing. The fuel is conveyed in a pneumatic system to a storage bin at the boiler plant. In the Ames project, the city-owned and operated processing plant produces the supplemental fuel for delivery to the electric generating plant operated by the city-owned Municipal Electric Service. Credit for the fuel is based upon the equivalent fuel cost saved by burning the supplement. The city also sells ferrous metals.

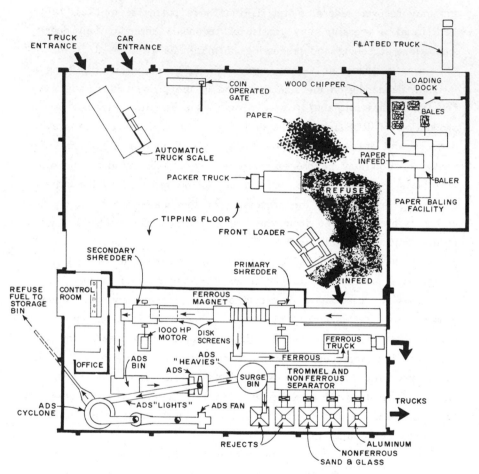

Figure 4.6.   Example Plant Layout for the Production of Suspension-fired RDF (Ames, Iowa)

Aluminum and mixed non-ferrous metals have been separated and sold only on a sporadic basis. Recent operating economics for the Ames plant have been reported (41). Net operating costs (net of RDF credits, recovered materials, and a minimal tipping fee) in 1978 averaged $12.50 per input ton.

Supplemental fuel, when burned in suspension, is burned at rates of up to 20 percent RDF to 80 percent coal on a Btu basis.

Other systems and unit processes are available to produce supplemental fuels for suspension firing. For example, Combustion Equipment Associates, Inc. (CEA), has developed a proprietary system to produce a powdered fuel, called Eco-Fuel II. This system produces a fuel similar to a dry process fuel, which is then chemically treated to embrittle the cellulose and reduced to a powder in a ball mill. The CEA plant in East Bridgewater, Massachusetts, which was producing an RDF for suspension firing has been modified to produce Eco-Fuel II. Another facility has been constructed in Bridgeport, Connecticut, which produces this type of fuel for co-firing with residual fuel oil in an existing boiler. Reference 42 contains a detailed description of the Bridgeport, Connecticut, plant. These plants are currently shut down pending resolution of the financial problems of CEA.

c. Densified Supplemental Fuel - The advantages of densified or pelletized supplemental fuel are 1) densified fuel will store for longer periods of time than shredded fuel before decomposing, and 2) the higher bulk density and shape of the densified fuel pellets allows this fuel to be handled with existing coal-handling equipment. Existing coal-handling equipment is normally not compatible with non-densified shredded fuel. Smaller, industrial boilers might be served better by such a fuel rather than the non-densified fuels described previously. Consequently, several companies now market systems or equipment which produce various forms of densified fuel. Densified RDF (d-RDF) is typically produced with size reduction to less than 1 inch nominal particle size, magnetic separation, air density separation and densification in various types of pelletizing or briquetting machines. Equipment of this type is currently installed in the Baltimore County facility operated by Teledyne National in Cockeysville, Maryland (43). A pilot scale d-RDF production plant at the National Center for Resource Recovery (NCRR) Equipment Test and Evaluation Facility in Washington, D.C. was operated on a trial basis in 1977 (44). The Black Clawson process (see section G, "Steam/Electric Systems") has also been the basis for d-RDF production at the Franklin, Ohio, plant.

There are no boilers currently burning d-RDF as a supplement to coal on a permanent basis. There have, however, been several test runs. Tests on traveling grate boilers at up to 25 percent (Btu basis) d-RDF were made in 1975 at Piqua, Ohio, and Wright Patterson AFB. Tests at up to 100 percent d-RDF were run on a vibrating grate boiler in 1976 at the University of Wisconsin Campus boiler plant. U.S. EPA sponsored test burns of d-RDF in 1977 at Hagerstown, Maryland, on a traveling grate type boiler have also been reported (45).

The results of both production and burning tests indicate that 1) the densification process is costly because of high wear rates on the densification equipment, 2) d-RDF can be stored for at least one month without problems, 3) existing coal handling equipment can be used to mix and transport d-RDF/coal mixtures without modification, and 4) d-RDF can be burned successfully in existing grate-fired boilers at up to 100 percent with minor operating adjustments.

2. Liquid and Gaseous Fuels

Liquid and gaseous fuels have been produced from solid waste by thermo-chemical reactions (pyrolysis), by bioconversion action, and by acid hydrolysis. These three major categories of systems which produce liquid and gaseous fuels are discussed in the following paragraphs.

a. Pyrolysis Systems - Pyrolysis systems thermo-chemically convert solid wastes into more usable fuels. Pyrolysis is a process whereby organic materials are heated in an oxygen deficient atmosphere to produce a gaseous or liquid product and a solid, carbon-rich residue. Man has been pyrolyzing coal for hundreds of years to produce coke and has been pyrolyzing wood for thousands of years to produce charcoal. Some decades ago, the process was adapted to produce petroleum coke. In recent years, there has been much publicity associated with adapting this age-old process to energy recovery from solid wastes.

The characterization of the liquid or gaseous products is generally a function of the pyrolysis reaction itself; primary bases for characterization are the temperature and speed of the reaction as well as the rapidity with which the pyrolysis products are quenched. Another factor affecting the characterization is the presence of other inert gas dilutants, such as nitrogen or carbon dioxide; however, since the feedstock material is predominantly cellulose, the pyrolytic products will be some form of a carbon-hydrogen-oxygen mixture. Generally, this will take the form of a methane-like product, some carbon monoxide, as well as some free hydrogen.

In all cases, the feedstock is fed into a pyrolytic converter or reactor, and heat is supplied either externally or through the internal exothermic reaction to drive the process. The pyrolytic gases or liquids which evolve are collected, drawn off of the reactor, quenched, and stored or combusted in a nearby boiler.

The primary advantage of a pyrolysis system is that the refuse can be transformed into gaseous or liquid fuel products which can be utilized by a wider variety of buyers than a solid refuse derived fuel product. However, both technological and economic feasibility have been major stumbling blocks for development of pyrolysis. Although the significantly modified City of Baltimore plant has been operated, its economic feasibility has been estimated as marginal at best (46,47). The plant is now closed during consideration of replacing the plant with a waterwall mass burn technology. No other full-scale pyrolysis plants are in operation.

The principal applications for pyrolysis systems are in those locations where environmental or market conditions dictate the exclusive use of a gaseous or liquid fuel product. A necessary prerequisite for this installation would be a thoroughly confirmed, reliable buyer for the pyrolysis products willing to pay the premium price necessary to produce this fuel. Table 4.2 provides information on the systems which have operated, or will soon operate, on a demonstration scale or larger.

TABLE 4.2

Large Scale Pyroysis Systems

Location	System Type	Products	Status as of 1/81[a]	Capacity (tons/day)
Baltimore, MD[b]	Formerly Monsanto Langard tm process, significantly modified	pyrolysis gas burned on site in a boiler to produce steam	Shut down in 1977, modified and re-opened in 1979, shut down for replacement in 1980	600
South Charleston, WV	Union Carbide Purox tm	pyrolysis gas	demonstration plant shut down in 1976	200
Orange County, FL	Andco-Torrax slagging Pyrolysis	pyrolysis gas burned on site to provide hot water for cooling and heating	under construction	100
San Diego County, CA	Occidental Petroleum, Flash Pyrolysis	pyrolysis oil	demonstration plant shut down after initial testing, modifications being considered	200
Luxembourg	Andco-Torrax slagging Pyrolysis	pyrolysis gas	In shakedown	200

a NCRR Bulletin, September 1980 (46) and GRCDA Report (43).
b See reference 47 for details of this system.

A pyrolysis process which can convert the organic portion of solid waste into gasoline has been demonstrated on a bench scale in a laboratory, but no full-scale plants of this type have been constructed (48).

b. Bioconversion Systems - Bioconversion systems utilize naturally-occurring microbes in the solid waste to biologically degrade the cellulosic fraction in an anaerobic environment. The biological action generates methane gas which can be burned as a fuel. Methane generation in landfills has traditionally been viewed as a problem. However, several methane collection systems at in-place landfills have been initiated in California, Illinois, New York, and New Jersey to utilize the methane as a fuel (46). The gas is typically collected from wells drilled into the landfill, and gas is withdrawn through underground collection systems. The gas must be purified and compressed before it can be used. Treatment to remove moisture, carbon dioxide, and other impurities must be performed to derive pipeline-quality gas.

Landfill gas recovery systems of this type have been built at the Palos Verdes Landfill (in operation since 1977), and the Sheldon-Arletta Landfill both near Los Angeles. Consult the reference materials (49,50) for further details. The general criteria for methane recovery in a landfill include a depth of approximately 40 feet, 2 million tons of refuse in place, a high percentage of organic waste, adequate moisture content, suitable cover material, and close proximity to an energy market.

Another form of bioconversion utilizes mechanical anaerobic digestion processes to recover methane gas. This process has been used for many years in wastewater treatment plants to digest the sludge and produce energy for plant operations. An experimental 50 to 100 ton per day "proof of concept" facility is operating in Pompano Beach, Florida. The facility, funded by the U.S. Energy Research and Development Administration, shreds, screens and air classifies waste, and blends the light fraction with sewage sludge, water and nutrients. The mixture then is anaerobically digested in mechanically agitated tanks for about five days. Recovered gases are about 50 percent methane and 50 percent carbon dioxide. Dewatered residue, equal to equal to about 30 percent of the volume of the raw refuse and sewage sludge will be incinerated or landfilled. Further technical details about mechanical bioconversion systems can be found in the reference materials (51,52).

  c. Acid Hydrolysis Systems - The acid hydrolysis of cellulose has been studied extensively for nearly a century. Most past research has concentrated on the conversion of wood wastes into ethanol. More recent research by the Department of Applied Science at New York University has resulted in the construction of a pilot-scale (1 ton per day) facility to convert newspapers into glucose which can then be used as a feedstock for the production of fuels (ethanol) by standard fermentation techniques (53). This pilot-scale plant utilizes hyropupling of the paper, irradiation pretreatment with an electron beam accelerator, and hydrolysis at 450 degrees F in a 3 percent solution of maleic acid in a continuous screw reactor. This research has shown glucose yields of up to 60 percent of input cellulose. Other researchers have reported on techniques to anaerobically digest the glucose from the acid hydrolysis reaction into methane gas (54).

  While this technology seems to hold promise for the future, no large-scale plants have been built, and the economic feasibility has yet to be proved. At this time acid hydrolysis must be categorized as "emerging technology" for resource recovery from solid waste.

G. Steam/Electric Systems

A method of energy recovery from solid waste which is now being practiced in this country, and which has been practiced in Europe for many years, is the production of steam by burning solid waste as a primary fuel in a new boiler. Auxiliary fossil fuels may be used as a supplement for light-off and flame stabilization. The steam can be used for industrial processes or can be utilized in a turbine-generator to produce electricity.

There are basically two types of steam plants which fire solid waste as the primary fuel: raw waste systems and dry or wet processed waste systems. These systems are explained in more detail in the following sections. Table 4.3 lists facilities in operation or under construction which use refuse as their primary fuel in the U.S. and Canada.

1. Raw Waste Systems

As the title indicates, this process involves feeding solid waste into a boiler without any mechanical preparation. This is also commonly referred to as "mass burning." The waste is stored in a pit, and is fed by crane to a moving grate type furnace. Bulky items are either excluded from the process or are shredded or sheared before being fed into the furnace. The ash from the furnace is quenched in a wet ash handling system, and is then often landfilled. Ferrous metals can be recovered from the ash, but are not as valuable as they would be if they were recovered before incineration. The stack gases are cleaned, and wastewater from the ash systems must be treated. Figure 4.7 shows a typical steam plant fired with raw waste. Figure 4.8 depicts a schematic flow chart for a raw waste fired system (with co-generation).

Smaller raw waste systems utilize modular incinerators. These combustion units have had wide usage as volume reduction systems for commercial waste, and are now finding a new application for municipal waste in conjunction with heat recovery equipment. The modular incinerator is available in sizes from 75 pounds to one ton per hour of solid waste. These units generally consist of a primary combustion chamber, secondary combustion chamber, and heat exchanger. The units are fed by a mechanical or hydraulic ram which charges the material onto a fixed bed. The waste is ignited by the use of auxiliary fuel until temperatures are reached to allow for autoignition. Combustion in the primary chamber is usually in a starved air atmosphere, often referred to by manufacturers as "pyrolytic incineration." The advantage of this starved air approach is that a high particulate exhaust is not generated. The combustible gases generated in the primary chamber

TABLE 4.3

Large Scale Facilities Firing Refuse As A Primary Fuel

Location	Type of Process	Status as of 1/81[a]	Design Capacity (tons/day)	Boiler Data
Akron, OH	Shredded air classified fuel; ferrous recovery	Operational	1000	travelling grate
Albany, NY	Shredded, air classified fuel, ferrous, nonferrous, recovery from boiler ash	in shakedown	750	travelling grate
Braintree, MA	Mass burn	Operational since 1971; metal recovery system planned	250	waterwall incinerator
Chicago, IL Northwest Incinerator	Mass burn; incinerated ferrous recovery	Operational since 1971; steam line completed 1980	1600	waterwall incinerator
Columbus, OH	shredded, air classified fuel, ferrous	under construction	2000	travelling grate
Dade County, FL	Wet processing; ferrous glass, non-ferrous recovery	In shakedown	3000	travelling grate
Duluth, MN	shredded fuel, and sewage sludge (co-disposal)	In shakedown	400	fluidized bed

Location	Type of Process	Status as of 1/81[a]	Design Capacity (tons/day)	Boiler Data
East Hamilton, Ont. SWARU	Shredded fuel; ferrous recovery	Operational since 1972	600	travelling grate
Glen Cove, NY	mass burn with sewage sludge	Under construction	225	waterwall incinerator
Hampton, VA NASA	Mass burn	In shakedown	200	waterwall incinerator
Harrisburg, PA	Mass burn, incinerated ferrous recovery, sludge drying	In operation since 1972; steam line completed in 1978	720	waterwall incinerator
Hempstead, NY	Wet processing; ferrous, glass, non-ferrous recovery	In shakedown, temporarily closed	2000	travelling grate
Montreal, Que.	Mass burn	Operational since 1970	1200	waterwall incinerator
Nashville, TN	Mass burn; incinerator ferrous recovery	Operational since 1974	720	waterwall incinerator
Niagara Falls, NY	Shredded, air classified fuel; ferrous recovery	In shakedown	2200	travelling grate

TABLE 4.3
continued

Large Scale Facilities Firing Refuse As A Primary Fuel

Location	Type of Process	Status as of 1/81[a]	Design Capacity (tons/day)	Boiler Data
Norfolk Naval Shipyard, VA	Mass burn	Operational since 1967	360	waterwall incinerator
Rochester, NY Kodak	Shredded fuel; silver recovery	Operational since 1970	180	tangential suspension-fired
Saugus, MA	Mass burn; incinerated ferrous recovery	Operational since 1975	1200	waterwall incinerator
Quebec City, Que.	Mass burn	Operational since 1974	1000	waterwall incinerator

a Source: NCRR Bulletin, September, 1980 (46), and GRCDA Report (43).

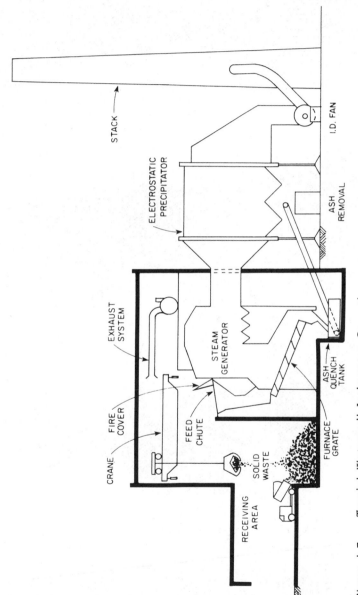

Figure 4.7. Typcial Waterwall Incinerator Operation

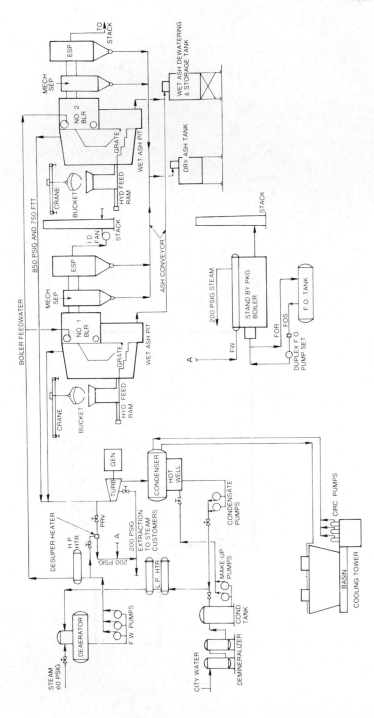

Figure 4.8.   Schematic Diagram of a Mass Burn Steam Plant

are completely combusted in the secondary chamber with the aid of
auxiliary fuel. The hot gases are then passed through a fire tube or
water tube boiler for heat exchange.

The advantages of these systems are their simplicity and their
application to small communities and other small waste generators.
The disadvantages of these systems are the poor energy recovery
efficiency and their inability to respond to varying types of waste.

There are about twenty installations of this type existing or under
construction in the U.S. (46). The success of these installations both
technically and economically has been mixed. A technical evaluation of
the Marysville, Ohio, and the North Little Rock, Arkansas, facilities has
been reported (55). The findings include a heat recovery rate of
between 42 and 56 percent of the available heat released, with
particulate emissions within local air quality standards.

2. Dry Processing Systems

This type of system involves shredding and magnetic separation of the
waste, and possibly air classification, before it is fired. The waste is
then stored, and is metered onto a traveling grate using a pneumatic or
mechanical conveyor system. This process can use either a wet or dry
ash handling system. Ferrous metals (and possibly other materials) are
recovered in the process. The stack gases must be cleaned prior to
discharge into the atmosphere.

The first plant of this type was built in 1972 in East Hamilton,
Ontario. The most recent facility of this type is located in Akron,
Ohio (see Table 4.3). The process utilizes two, 60 ton-per-hour
shredders followed by air density separation. Ferrous and non-ferrous
metals are removed from the "heavy" stream and the "light" fraction is
stored in a traveling screw-type bin. The boilers adapted from the
bark burning industry are of the traveling grate type with spreader
stokers in which much of the combustion will occur in suspension. An
electrostatic precipitator is used for air pollution control. Steam is

sold to a district heating system and used to produce in-house electricity. Reference 56 contains a detailed description of the Akron system.

3. Wet Processing Systems

As an alternative to dry solid waste processing systems, the Black Clawson Company has developed a patented "Hydrasposal System" based on paper making technology to recover paper fibers from municipal solid waste. Because the wet pulping process was restricted to regions in close proximity to paper fiber markets, Black Clawson shifted its emphasis to the recovery of a fibrous fuel to compete with other forms of refuse derived fuel. The wet-pulped fuel can be produced in two grades; the No. 1 product is a cone-pressed material with a 50 percent moisture content while the No. 2 product is dried to a 5 t0 20 percent moisture content and may be pulverized or pelletized. The No. 1 product is used as the primary fuel for firing in specially designed bark furnaces for effective combustion. The No. 2 fuel, however, can be used as a supplemental fuel in existing boilers. The energy efficiency of the Black Clawson No. 2 process is reduced by the need for drying.

The wet separation process had been demonstrated on the scale of 150 tons per day (TPD) in Franklin, Ohio, from 1972 to 1979. Based on this technology, Hempstead, New York has constructed a system and Dade County, Florida, has recently completed construction of the Black Clawson System. The No. 1 fuel product generated will be combusted in a bark furnace for electrical power generation. The recovery system will also reclaim ferrous metal, aluminum and glass products. The Hempstead project has an estimated project cost of $81 million. A twenty-year contract has been signed with Long Island Lighting Company, to purchase the generated electricity. Five-year contracts have been signed for the secondary materials. The contracted drop charge of $15.00 per ton for solid waste disposal is competitive with other disposal alternatives in the region. The plant has temporarily been shut down due to contractural disputes and emissions problems. The Dade County project has an estimated project cost of $165 million and is similar in scope to the Hempstead project.

4. Co-generation

"Co-generation" is a new term for an old and proven technique of energy production. The term "co-generation" in its broadest definition means any simultaneous production of electric power (or mechanical energy) and useful thermal energy (steam or hot gases). Co-generation has come to include a number of processes, some of which are not applicable to solid waste resource recovery:

1. Boiler/Steam Turbine systems which either extract steam for sale or utilize back-pressure turbines while producing electric power

2. diesel engines driving generators with waste heat recovery

3. gas turbines driving generators with waste heat recovery

4. other energy production systems utilizing Sterling engines, fuel cells, and thermionic devices with waste heat recovery

Co-generation systems have been traditionally divided into two fundamental types:

Topping Systems - electricity is produced first, and the exhausted thermal energy (usually in the form of steam or hot gases) is put to further use.

Bottoming Systems - thermal energy is produced (usually in the form of steam) and first used in an industrial or commercial process, then still-usable thermal energy is extracted for further use.

Solid waste energy recovery co-generation systems are limited almost exclusively to boiler/steam turbine topping systems. Internal combustion engines and gas turbines have not as yet shown the technical feasibility for firing solid waste derived fuels with the possible exception of methane recovered from landfills. The fossil fuel fired steam boiler is the only combustion system proven in actual commercial-scale operation with solid waste derived fuels. Although a bottoming boiler/steam turbine co-generation system utilizing solid waste as a fuel is technically feasible, industrial buyers of refuse produced steam have not indicated widespread interest in this concept. Solid waste co-generation systems typically involve the production of high pressure superheated steam which is used to turn a turbo-generator unit to produce electric power. Steam at a lower pressure is sold, usually to a large industrial steam user.

There are two basic approaches for solid waste energy recovery co-generation. The "extraction approach" and "back-pressure approach" are illustrated in Figures 4.9 and 4.10. The choice between the two approaches depends on the steam buyer's demand versus the fixed amount of steam which can be produced from available solid waste. The extraction approach is most applicable in situations in which the selected steam buyer has a steam demand less than the amount available from the solid waste over a significant portion of the year. An extraction-type turbine is used in which steam is pulled off of the turbine at the desired pressure through a special port. This extracted flow can be controlled to meet demand variations. Steam not extracted continues through the turbine to the condenser producing additional electricity.

Co-generation may also be applicable in situations in which the steam buyer always has a steam demand equal to, or greater than, the amount available from the solid waste over the entire year. In this situation, the back-pressure approach can be used. Steam is generated at high pressure, superheated conditions, and used to turn a back-pressure turbogenerator unit. The "back-pressure" term is used because the turbine exhausts steam for sale to the steam buyer at the required pressure (typically 120-200 psig) rather than at a typical condensing pressure (1.5 to 2.5 in. mercury). In this situation, the steam buyer utilizes solid waste generated steam only as a supplement to the buyer's existing on-site boilers.

The selection and operation of either co-generation approach is also influenced by the selection of either the steam buyer or the electric power buyer as the "primary market." A primary market must be selected because the sale of "firm" steam to a steam buyer with the

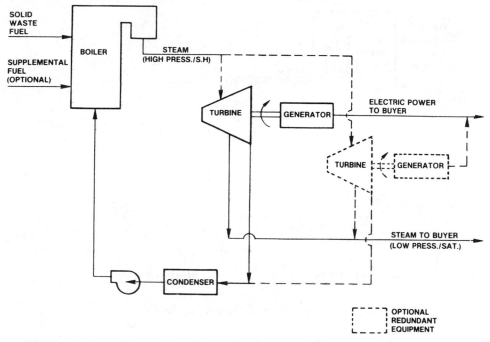

SOLID
WASTE
FUEL

SUPPLEMENTAL
FUEL
(OPTIONAL)

BOILER

STEAM
(HIGH PRESS./S.H)

TURBINE

GENERATOR

ELECTRIC POWER
TO BUYER

TURBINE — GENERATOR

STEAM TO BUYER
(LOW PRESS./SAT.)

CONDENSER

OPTIONAL
REDUNDANT
EQUIPMENT

Figure 4.9.   Co-Generation: Extraction Approach

generation of electric power from the excess steam implies different
operation (and in some cases different equipment) than the sale of
"firm" electric power with the sale of any excess steam.   If the steam
buyer is the primary market, either approach is applicable depending on
steam demand versus available steam generation.   If the extraction
approach is used with a primary steam market, equipment redundancy
may be required because the co-generation plant must provide the full
steam buyer's demand at all times.   If the electric power buyer is
selected as the primary market, the extraction approach is more
flexible because the extraction flow can be throttled to control electric
power generation variations.

Co-generation with solid waste fired steam boilers may be an
economical energy recovery technique if there exists a willing electric
power buyer (either an industry or a utility) and a willing steam buyer.
Market values for steam and electric power are critical to the
economic feasibility of co-generation versus electric-only, or steam-only

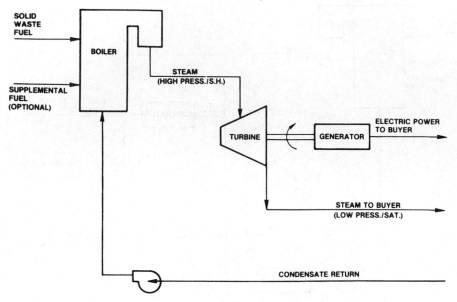

Figure 4.10.   Co-Generation: Back-Pressure Approach

systems.   The advantages of co-generation for solid waste energy recovery are summarized below.

1.   Co-generation is energy efficient.   This type of facility can produce steam and electric power with 10 percent to 30 percent less fuel than a combination of similarly sized steam-only and electric-only systems.

2.   The environmental impacts of higher efficiency energy production are less.

3.   If market and regulatory conditions are favorable, co-generation with solid waste fired equipment can have an economic advantage over steam-only or electric-only systems.

5.   Co-Disposal

Co-disposal refers to the combined disposal of wastewater treatment plant sludge and municipal solid waste.   Wastewater sludge contains a very high percentage of moisture, commonly 95 percent to 98 percent before concentration. The moisture content is dependent upon the wastewater treatment process from which it is generated. The various sludge concentration processes can reduce the moisture content of sludge to 70 percent to 85 percent. This reduction in moisture content is very expensive, requiring elaborate equipment and using considerable electrical energy.

Typical municipal sewage sludge is generally autogenous at 70 percent moisture (or 30 percent solids). At this point the entire heating value of the sludge solids is required to vaporize the remaining moisture. Approximately 1100 Btu of heat energy is required to drive off every pound of water in the sludge. It is evident, therefore, that sewage sludge is not a fuel. If the expensive technique of drying is used to dewater the sludge to the point where it can yield heat energy above that required to incinerate itself, then it has some value as a fuel. The true value of co-disposal is that it offsets the costs incurred for disposal of sludge, thereby encouraging resource recovery.

Co-disposal has been tried numerous times in the recent past through incineration. Municipal incineration of solid wastes (combustion as a disposal method, not for energy conversion) was popular in the U.S. from the late 1940's through the 1960's. During this time various methods of feeding sludge were tested. These methods included pre-drying of the sludge and feeding in combination with refuse, spraying refuse with sludge before feeding, and spraying sludge directly into the incinerator. The use of refractory-lined refuse incinerators in this country has been generally discontinued because of operating costs and air pollution regulations.

The most recent co-disposal developments include combustion of a solid waste and sewage sludge mix in various types of boilers and the bioconversion of a solid waste and sewage sludge mix to produce methane. Combustion co-disposal systems have included utilizing waterwall incinerators, multiple hearth furnaces, and fluidized bed reactors. Two facilities in France and Germany have been co-disposing of sewage sludge and solid waste in waterwall incinerators for several years. A similar system is now under construction in Glen Cove, New York. The multiple hearth furnace concept has been extensively tested in Contra Costra County, California and there are plans to construct a

facility to burn 1000 tons per day of solid waste along with the sludge
from a 45 million gallons per day (MGD) wastewater treatment plant.
A fluidized bed co-disposal system is currently in shakedown in Duluth,
Minnesota.   The Duluth system burns about 200 tons per day of
shredded, magnetically separated, air classified solid waste along with
340 tons per day of sewage sludge with 20 percent solids (57).   A
bioconversion co-disposal facility in Pompano Beach, Florida, has been
previously described.

III. SELECTING SYSTEM ALTERNATIVES

To select alternative systems for economic analysis, the investigator
must specify the primary facility, the landfill(s) which will be used, and
the transport system for each alternative.   The set of alternatives must
be properly sized, the chosen technologies must be compatible with
market requirements, and the alternatives must be comparable both
with respect to the amount of waste handled and the time period.   The
following sections give guidelines on how to properly define system
alternatives.

A. Primary Facility

The primary facility in an alternative system may be a landfill or a
variety of different resource recovery facilities.

1. The Landfill Alternative

A landfilling alternative should always be included for use as a baseline
comparison with resource recovery.   If the presently used landfill or
landfills have at least twenty years' capacity, then the present system
can be used as the landfill alternative.   If multiple landfills are

presently used, the investigator should be certain to consider the increased fill rates at certain landfills which will result from the completion and closure over time of other landfills when calculating remaining capacity. Discussions with the operators (municipal or private) of the existing landfills should be held to determine total remaining capacity.

If the present system of landfills does not have adequate capacity to dispose of all solid waste in the study area over the period of analysis of the resource recovery facility (20 to 25 years), then a new landfill size and location must be postulated. Methods for calculating required landfill area for a particular study area have been discussed in Chapter 2. Methods for selecting potential sites have also been discussed. The critical factor in choosing the facility in the landfill alternative is that the landfilling system chosen must have the capacity for disposing of all or the same portion of the waste (processable and non-processable) over the same period as the period of analysis of the resource recovery alternatives. Otherwise, the landfilling alternative will not be comparable to the resource recovery alternatives.

2. The Resource Recovery Alternatives

The technologies selected for the resource recovery facilities must produce energy forms and materials for which there is a market. Further, it is the market for energy which will determine the selection of the basic technology. Energy sales comprise 95 percent of the revenue in a typcial resource recovery system, therefore it is essential that the selected technology meet the requirements of the energy buyer as a primary goal. Most marketable forms of materials can be separated in a variety of different energy recovery technologies, therefore, the selection of materials recovery equipment should be made after selection of the energy recovery technology.

The investigator must determine conceptual design guidelines for the resource recovery facilities which must be applied consistently between alternatives. Factors to be considered are covered in the following paragraphs.

a. Sizing - There are two approaches to the sizing of a resource recovery facility: by the waste stream quantity, and by the energy market demand. The selection of a particular approach depends on the size of the energy market demand versus the available energy in the solid waste and the basic philosophy of the economic analysis. The following three cases illustrate many of the sizing considerations which may be encountered.

i.   Case 1--waste limited. In a typical study area the amount of waste generated will vary seasonally with quantity peaks in the summer months. Consequently, the amount of available recovered energy varies porportionally. If the average energy demand is always greater than the available energy in the solid waste, the investigator must decide whether to size the facility for only the solid waste quantity, or if a larger facility using the addition of a conventional fuel to meet the higher average demand should be considered (see Figure 4.11, Case 1). Obviously, the larger facility should only be considered if the extra energy revenues offset the extra operating costs, debt service, and fuel costs. The investigator should consider that many times an energy buyer will pay a higher unit price for the replacement of the entire demand rather than a portion of it because the buyer can shut down all existing equipment. The investigator should also determine whether the municipality is interested in becoming a primary producer of energy for a certain customer in addition to building a system to handle solid waste disposal. Case 1 actually illustrates the ideal solid waste/market relationship because the entire output of the resource recovery plant can be purchased at all times of the year.

ii. Case 2--energy market limited. If the amount of energy available from the solid waste is more than the demand of a particular energy market at all times of the year, the resource recovery facility can be sized either to meet the energy market demand or the co-generation concept can be used to market excess energy from a larger facility sized for the waste generation (see Figure 4.11, Case 2). With a facility sized to meet the market demand, excess waste must be

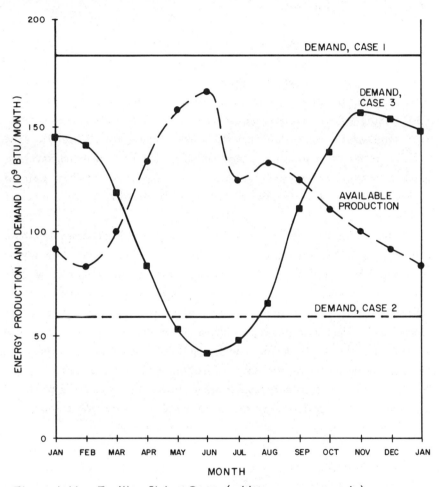

Figure 4.11. Facility Sizing Cases (arbitrary energy scale)

landfilled.   The facility can be designed for expansion so if another
energy market develops in the future, the excess waste can be shifted
from the landfill to an expanded facility.   The co-generation concept
has become particularly attractive with incentives contained in the
recent PURPA legislation (see Chapter 3).

        iii. Case 3--seasonal demand.   Many potential energy buyers
have a seasonal energy demand--particularly industries in northern
climates which use steam for space heating in the winter, and district
heating systems.   These energy markets can cause resource recovery
plant sizing problems because demand variations usually are opposite
from solid waste production variations (see Figure 4.11, Case 3).   The
resource recovery plant can either be sized for peak demand, average
demand, or peak solid waste generation.   Trade-offs between capital
costs, auxiliary fuel costs, and excess landfilling costs must be
considered.

        If the facility is to be sized for the amount of processable waste
available (as in Case 1), the investigator must decide what design year
to choose since in most study areas the waste quantity will grow over
time.   Other public utilities such as wastewater treatment plants, water
supply systems, and electric power systems are normally sized for
"worst case" conditions (maximum demand) so that adequate capacity is
ensured over the study period.   In sizing these facilities, therefore, the
designers usually choose the last projection year as the design year.
The sizing of a resource recovery plant should, however be approached
differently.   The design year for a resource recovery plant should be
the projected start-up year for the following reasons:

1.   Solid waste quantity projections are speculative.   The uncertainty
     in the validity of projections increases in direct proportion to the
     number of years projected.   No matter how diligently the
     investigator defines unit waste factors, the validity of the waste
     quantity projections depends on the validity of the population and
     employment projection models used for projection calculations.   It
     is much safer to size the facility for a waste quantity which is
     the least uncertain.

2. Unlike a wastewater treatment plant, a water supply system, or an electric power plant, a resource recovery plant is generally not the only facility available. The result of an underestimate in future solid waste quantity, therefore, is not catastrophic. An undersized facility simply means that excess waste must be diverted to a landfill.

3. The economic consequences of oversizing a resource recovery facility are severe. These facilities are characterized by high fixed costs such as debt service and certain operating and maintenance costs. Roughly half of the annual costs of a typical resource recovery plant are fixed. If less than the design tonnage is received, the burden of the fixed costs on underutilized equipment can raise per ton operating costs to unacceptable levels. Provisions for future expansion should, however be made in the design.

b. Technology - The technology selected for the resource recovery facility must first meet the requirements of the energy market. The investigator must know the energy form required (fuel, steam, electric power) delivery rates desired, and physical specifications (in the case of substitute fuel production). With this knowledge, the required equipment can be selected and sized. It is important that the investigator choose technologies which have been proven in full-scale operation over a significant portion of time. A proved technology is important not only from a reliability and cost standpoint, but also because financing an unproven technology will be difficult, if not impossible.

At this point in the investigation, the configuration of the facility must be determined in general terms. The type of receiving arrangement (floor dump or pit) and raw waste storage sizing must be determined. The number and capacity of processing lines, boilers and other equipment must be determined. The type of energy delivery system must also be detailed (pipeline, truck transport, electric power lines, etc.).

Either type of receiving configuration is acceptable; however, most processed waste systems use a floor dump configuration, and most mass burn facilities utilize the pit storage system with overhead crane retrieval. In the processed waste case, the raw waste storage area should be sized for at least one to two days of receiving without processing. For a mass burn facility, raw waste storage should be sized to accommodate enough waste for at least three full days of burning if the boilers are to operate continuously through three-day weekends.

The design of the processing line, if used, depends on the boiler or other energy conversion process requirements. The processing line number and capacity sizing depends on the redundancy required for on-line reliability and the number of daily shifts of operation desired. The capacity of typical processing lines is usually limited by the initial sizing equipment (either a shredder or a primary trommel). Shredders are available with capacities ranging from 15 to 100 tons per hour and primary trommels have been operated in the 50 to 85 tons per hour range. These, however are instantaneous rates, so the investigator should allow for a lower continuous rate due to short, unscheduled down-time periods which occur in normal operation. A factor of 60 percent to 80 percent should be applied to account for this downtime. Calculations can be based on a one- or two-shift daily operating schedule. Fewer processing lines will be necessary with a two-shift operation, but higher labor costs will result. A three-shift (24 hours per day) operation is not recommended for processing equipment because it is necessary to utilize at least one shift for equipment maintenance and housekeeping.

The selection of the numbers and sizes of boilers (if used) depends on the facility sizing criterion (solid waste quantity or energy market demand), the type of facility (mass burn or processed fuel), and the amount of redundancy required. Large-scale waterwall incinerators and traveling grate type boilers for processed refuse are available usually in the size range of 50,000 to 300,000 pounds per hour of steam per unit (150 to 1200 tons per day of solid waste). The maximum size of a traveling grate unit is limited to about 300,000 pounds of steam per hour because of size limitations on the grates and because grate heat release rates are limited to 200,000 to 1 million Btu/square foot (58).

Smaller modular incinerators with heat recovery are available in the approximate size range of 2,000 to 15,000 pounds per hour of steam per unit (5 to 25 tons per day of solid waste). If high steam delivery reliability is required by the energy market, the investigator should select multiple units sized so that at least one unit can be shut down for repairs while the others can provide the full steam flow. Large-scale waterwall incinerators and processed waste boilers can normally be expected to be on-line for 70 to 75 percent of the available annual hours (this accounts for both scheduled and unscheduled down-time).

The energy delivery system design depends on the form of energy produced. In the case of RDF, pneumatic or mechanical conveyors can be used if the boiler to be used is within 2,000 to 3,000 feet of the processing plant. If the boilers are more remote, truck transport utilizing vehicles similar to raw refuse transfer trailers can be used. RDF requires surge storage which should be sized to meet energy market demands during anticipated scheduled processing plant downtime. Gaseous fuels and steam must utilize pipeline delivery systems. Electric power must use an electric power distribution system, but the investigator should determine where the system tie-in point should be. This question can be answered by the utility purchasing the power.

c. Location - The location of the resource recovery facility depends on site availability and the type of energy being produced. For facilities producing solid fuels the location should be as close to the generation of the raw waste as possible. This is because raw waste transport costs are much higher than fuel transport costs. If the facility produces gaseous fuel or steam, it is necessary to locate the facility close to the energy buyer because of the high costs of pipelines. For facilities generating electric power, the facility should be located as close as possible to the areas of waste generation since a variety of locations can be used for the tie-in of electric power to the utility grid. As previously discussed, some systems may utilize a processing plant in one location, with the energy recovery facility at another location.

B. Landfills

Supporting landfills must always be a component of any resource recovery system alternative. A landfill is required for disposal of all non-processable wastes (which will be transported directly to the landfill), processing residues, and boiler or incinerator ash. Table 4.4 lists factors for rough calculations of processing residue and ash for several different resource recovery technologies. The amount of non-processable waste generated in the study area is available from local landfill data and methods for determining these quantities have been discussed in Chapter 2. Landfill space for excess processable waste may also be necessary in the situation in which the energy market demand dictates a resource recovery facility size smaller than the available processable waste generation.

Calculation methods for determining the acreage required for a given waste generation rate have been discussed in Chapter 2. These calculations should be made for the quantities to be landfilled under each alternative system. Existing landfills can be utilized if remaining capacity at the lower refuse delivery rates under a resource recovery alternative are sufficient for the entire study period (20 to 25 years). If not, new landfill space must be located as discussed in Chapter 2.

C. Transport System

In selecting the transport system to be used with each alternative system, the investigator must first decide if transfer stations should be considered. If the longest distance between a waste generation zone and a potential resource recovery facility or landfill location is 10 to 15 miles or greater, transfer stations should be considered. In general, study areas geographically smaller than this will not justify transfer stations.

If the decision is to consider transfer stations in the alternatives, the investigator must next decide which type of transfer station will be

TABLE 4.4

Typical Residue and Ash Quantity as a Percent of Raw Waste

System Type	Processing Residue	Fly Ash and Bottom Ash
Mass Burn[a]	0	35%
Mass Burn with Ferrous recovery from Ash[b]	0	27%
Processed Waste as primary fuel in dedicated boiler (ferrous recovered)[c]	22.9%	12%
Processed Waste for supplemental firing in suspension fired boiler (ferrous and aluminum recovery)[d]	22.3%	8%
Pyrolysis[e]	22.9%	12%

a   25% bottom ash, 10% fly ash and APC solids.
b   8.6% ferrous content, 90% recovery.
c   78% light fraction, 8.6% ferrous at 90% recovery RDF ash: 15%
d   78% light fraction, 8.6% ferrous at 90% recovery, 0.7% aluminum at 80% recovery, RDF ash: 10%.
e   Assumes primary processing as in note c and slag content equal to RDF ash content.

utilized. The two basic types of transfer stations (compacted and non-compacted) were described previously. The selection of either the compacted or non-compacted design is a function of the load limitations on the public roads in the study area and other factors.

A typcial 75 cubic yard compacted transfer trailer and tractor weighs about 42,000 pounds. If the gross vehicle weight (GVW) is limited to about 80,000 pounds, this truck can haul 19 tons of waste if a 500 pounds/cubic yard compaction factor is achieved. A typical 112 cubic yard non-compacted transfer trailer and tractor weighs about 33,000 pounds. If the gross vehicle weight is limited to 80,000 pounds, this truck can haul 23 1/2 tons of waste (with a compaction factor of about 416 pounds/cubic yard), or about 24 percent more waste than a compacted trailer. An uncompacted trailer weighs much less than a compacted trailer because of different unloading mechanisms and the fact that a compacted trailer must be heavily reinforced to withstand the compaction forces. Because of the payload size differential, an uncompacted trailer may be able to haul waste at a lower cost in states which have the 80,000 pound (or lower) gross vehicle weight limitation. However, additional equipment at the unloading site may be necessary with some uncompacted trailer designs. These cost trade-offs will factor into the selection.

The proper size of the transfer stations selected for inclusion in the alternatives cannot be determined exactly at this point in the analysis. The proper size depends upon how many waste generation districts will show an economic advantage in bringing waste to the transfer station rather than direct hauling to a landfill or resource recovery facility. Chapter 6 presents a mathematical model for determining transfer station sizing. It is necessary, however, for the investigator to make estimates of transfer station sizing for use in the mathematical model. By examining the potential locations for the transfer stations in relation to the locations chosen for the primary facilities (see Chapter 2), the investigator can visually estimate which WGDs might be closer to the transfer stations than the primary facilities. Summing the waste generation quantities in these WGDs will landfill of this capacity are about 640 acres for the land near River give an upper limit to the transfer station size which should be converted to units of tons per week. Two to three smaller transfer station capacities should be chosen for possible selection in the mathematical modeling.

IV. RIVER CITY ALTERNATIVES

The objective of the River City study is to provide a reliable, long-term, cost-effective system for managing solid waste for the citizens of River City. As such, the following guidelines were established prior to selecting system alternatives:

1. All alternative systems were sized to dispose of all processable and non-processable solid waste from 1985 through the year 2005. Facilities to serve energy markets with demands higher than the energy in the solid waste were sized for the available waste quantity rather than for full energy replacement with supplementary conventional fuel.

2. In order to provide reliable solid waste disposal service, only technologies which have been constructed and operated on a full-scale basis successfully were considered.

3. Sewage sludge co-disposal was not considered because current sludge disposal costs are minimal due to the sale and/or landspreading of the sludge.

With these guidelines in mind, and with the market information given in Chapter 3, the four alternative systems described in the following paragraphs were selected.

A. Alternative 1--New Landfill

Calculations show that landfill space for about 9 million tons of waste will be required for a new landfill. Land requirements for a new

City (mostly flat, compatible soils). Both the North Landfill expansion (LF1) and the new private landfill to the south (LF2) are candidate locations for this new landfill.

Visual estimation of potential transfer station sizes led to the selection of three candidate sizes: 1,250 tons per week (TPW) or 250 tons per day (TPD) on a 5-day basis; 2,500 TPW or 500 TPD; and 3,750 TPW or 750 TPD. A compacted trailer design was selected.

The conceptual design for the new landfill includes scales, scale house, maintenance building, and paved entrance roads. Ground water monitoring wells, a leachate collection system, and a leachate treatment system are also included. Operating equipment including scrapers, dozers, graders, water trucks, and other rolling stock are also included for operation at a rate of about 7,700 tons per week (about 400,000 tons per year to approximately match the projected annual quantity in 1985).

B. Alternative 2--Steam to RCI

River City Industries (RCI), located in the East Side Industrial Park is a potential steam market (see Chapter 3). RCI has an average steam demand of about 300,000 pounds per hour (150 psig, saturated), on a six day per week basis. The steam demand is essentially constant over all seasons because the majority of the steam is used as a heat source in the industrial process rather than for space heating or cooling.

The mass burn technology was selected to serve this market on a total steam load replacement basis. Figure 4.12 is a schematic diagram of the steam plant. The plant will include scales, pit storage, crane retrieval, two 200,000 pounds per hour (steam), 800 tons per day (solid waste) mass-fired waterwall incinerators (60 psig, saturated steam conditions), a 100,000 pounds per hour (steam) gas/oil fired back-up boiler, and all necessary buildings and other equipment. Assuming an annual average boiler availability of 75 percent, the design capacity is

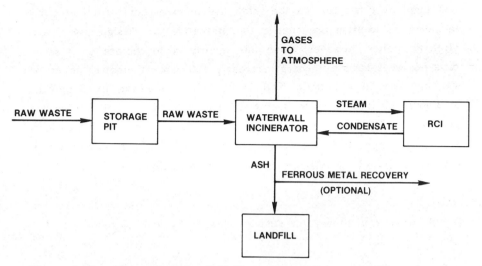

Figure 4.12.   Alternative 2:   RCI Steam Plant Schematic Diagram

about 374,000 tons of solid waste per year if operated on a six days per week basis (7,200 TPW). A location in the East Side Industrial Park is assumed, and a 2,500 ft. steam line and condensate return line to the RCI plant are included. Magnetic separation of the ash to recover ferrous metals is also included in the processs.

Transfer station potential locations, design, and size ranges as previously discussed for Alternative 1 are also assumed.

C.   Alternative 3--Co-Generation

Since there are markets for both steam (RCI) and electric power (RCPL), a co-generation technology was selected as an alternative. The selected combustion technology is similar to Alternative 2 (mass burn), but generates steam at 600 psig, 750 degrees F in two waterwall incinerators, each rated at 188,000 pounds per hour (pph) of steam (800 TPD). The steam is introduced into an extraction type turbine in which 150 psig steam is extracted for sale to RCI, and throttled to meet their demand. The electric power produced is sold to River City Power

and Light by connection to the grid. All other aspects of this facility
including its location are similar to Alternative 2. Design capacity is
slightly higher, however, since this facility could operate on a seven
days per week basis by generating nearly 100 percent electric power on
the one day in the week RCI is closed. Assuming a 75 percent
availability factor, design capacity is about 440,000 tons of solid waste
per year (8400 TPW). Also, transfer station design, potential locations,
and size ranges are the same as the first two alternatives. A
schematic diagram of this alternative is given in Figure 4.13.

D. Alternative 4--Steam For District Heating System

Even though the existing River City District Heating System (RCDHS)
is a potential market for pyrolysis or bioconversion gas, it was decided
that neither of these technologies has yet to demonstrate long-term
full-scale technical feasbility. The production of steam to serve this
market was, therefore, the only technology considered.

Because of the downtown location of the existing RCDHS steam
plant, there is insufficient space at or near the site for a waterwall
incineration facility similar to Alternatives 2 and 3. The construction
of a steam line from a remote steam plant would be prohibitively
expensive. The system chosen for this alternative, therefore, includes a
waste processing plant located at the existing transfer station site (TS
4) near the downtown area, and two processed waste traveling grate
type boilers located near the existing RCDHS steam plant. The
existing transfer station would be modified to include equipment which
produces RDF for truck haul about one mile to the two boilers.

The proposed modifications to the transfer station are shown
schematically in Figure 4.14. The processing equipment includes two
identical, 75 tons per hour processing lines. Each line includes a
primary trommel screen, shredding of the primary trommel overs,
magnetic separation, a secondary trommel, and air classification.
Stationary compactors are used to load the RDF into compacted

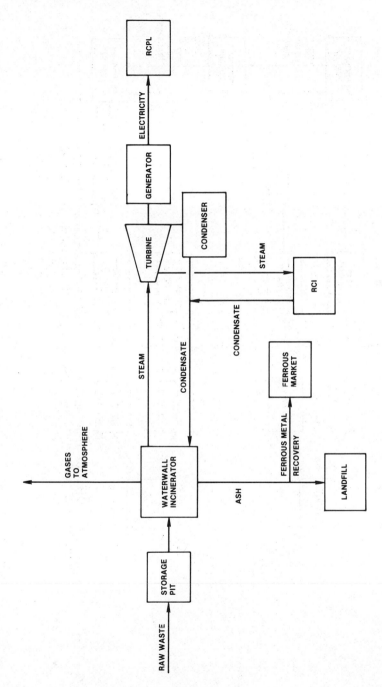

Figure 4.13.   Alternative 3:   Co-Generation Schematic Diagram

180

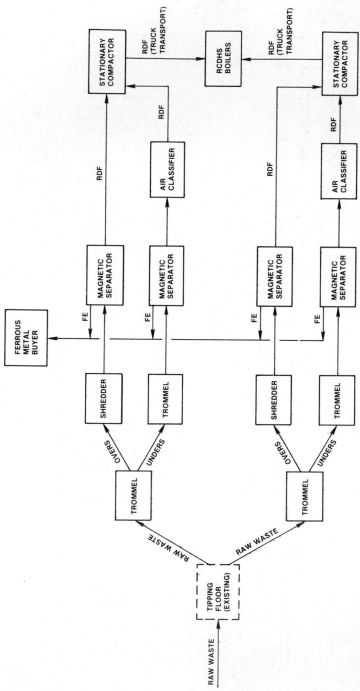

Figure 4.14. Alternative 4: District Heating System Steam Plant Schematic Diagram

transfer trailers for haul to the downtown steam plant. No separation systems for aluminum or glass are included; however, space for the future addition of aluminum separation equipment has been included. The RDF produced would have a nominal particle size of 3 inches. Operation of the processing plant would be on a two-shift per day basis, six days per week. Design capacity assuming a 60 percent availability factor on the processing lines is about 450,000 tons of solid waste per year (about 8650 TPW).

The boilers installed near the existing steam plant would be both rated at 200,000 pph. This boiler size was selected so that present peak steam demand (360,000 pph) could be met. Future increases in demand would be met with existing units at the RCDHS plant. Steam conditions demanded by RCDHS are 650 psig at 750 degrees F because RCDHS wishes to continue to use existing turbine-driven chillers for chilled water production. Also included in this installation would be an RDF receiving building, a surge storage bin, mechanical RDF conveyors to the boilers and other auxillary boiler equipment. The boilers would be operated on a 24-hour per day basis seven days per week. No redundant units or back-up gas/oil fired units are required since this equipment already exists in the present RCDHS steam plant. The existing boilers would be operated during periods of downtime for the RDF boilers.

Chapter 5 presents a cost analysis for these Alternatives.

REFERENCES

1. U.S. Environmental Protection Agency, Criteria for Classification of Solid Waste Disposal Facilities and Practices, Federal Register, Vol. 44, No. 179, Sept. 13, 1979, p. 53438.

2. U.S. Environmental Protection Agency, Classifying Solid Waste Disposal Facilities - A Guidance Manual, SW-828, March, 1980.

3. U.S. Environmental Protection Agency, Sanitary Landfill Design and Operation, D.R. Brunner, and D.J. Keller, SW-65ts, 1972.

4. U.S. Environmental Protection Agency, <u>Decision-Makers Guide in Solid Waste Management</u>, SW-500, 1976.

5. Ralph Stone, Sanitary Landfill, in <u>Handbook of Solid Waste Management</u>, D.G. Wilson ed., Van Nostrand Reinhold, New York, 1977, p. 226.

6. R. Hering and S.A. Greely, <u>Collection and Disposal of Municipal Refuse</u>, McGraw-Hill, New York, 1921.

7. Ronald J. Alvarex, Status of Incineration and Generation of Energy from Thermal Processing of MSW, in <u>Proc. 1980 National Waste Processing Conf.</u>, ASME, May 11-14, 1980, p. 211.

8. J.L. Pavoni, et al., <u>Handbook of Solid Waste Disposal</u>, Van Nostrand Reinhold, New York, 1975.

9. Adel F. Sarofim, Thermal Processing: Incineration and Pyrolysis, in <u>Handbook of Solid Waste Management</u>, D.G. Wilson ed., Van Nostrand Reinhold, New York, 1977, p. 166.

10. N.J. Weinstein and R.F. Toro, <u>Thermal Processing of Municipal Solid Waste for Resource and Energy Recovery</u>, Ann Arbor Science Pub., Inc., Ann Arbor, 1976.

11. U.S. Environmental Protection Agency, <u>Size-Reduction Equipment for Municipal Solid Waste</u>, SW-53c, Midwest Research Institute, NTIS, No. PB-226 551, 1974.

12. U.S. Environmental Protection Agency, <u>Solid Waste Milling and Disposal on Land Without Cover</u>, Vols. 1 and 2, J.J. Reinhardt and R.K. Ham, NTIS Nos. PB-234 930 and PB-234 931, 1974.

13. U.S. Environmental Protection Agency, <u>High-Pressure Baling of Solid Waste</u>, SW-32d, K.W. Wolf and C.H. Sosnovsky, 1972.

14. K.W. Wolf and C.H. Sosnovsky, "High-Pressure Baling and Compaction of Solid Waste," in <u>Handbook of Solid Waste Management</u>, D.G. Wilson ed., Van Nostrand Reinhold, New York, 1977, p. 136.

15. U.S. Environmental Protection Agency, <u>Solid Waste Transfer Stations: A State-of-the-Art Report on Systems Incorporating Highway Transportation</u>, SW-99, T. T. Hegdahl, NTIS No. PB-231 511, 1972.

16. U.S. Environmental Protection Agency, <u>Recovery and Utilization of Municipal Solid Waste</u>, SW-10, N.L. Drobney, H.E. Hull, and R.F. Testin, 1971.

17. W.J. Johnson, Shredding of Solid Wastes, in <u>Handbook of Solid Waste Management</u>, D.G. Wilson ed., Van Nostrand Reinhold, New York, 1977, p. 150.

18. Arthur F. Taggart, Handbook of Mineral Dressing, John Wiley and Sons, Inc., New York, 1927.

19. U.S. Environmental Protection Agency, Evaluation of the Ames Solid Waste Recovery System, Part 1: Summary of Environmental Emissions; Equipment, Facilities and Economic Evaluations, EPA-600/2-77-205, Nov. 1977.

20. Douglas E. Fiscus et al., Evaluation of the Performance of the Disc Screens Installed at the City of Ames, Iowa, Resource Recovery Facility, In Proc. 1980 National Waste Processing Conf. ASME, May 11-14, 1980, p. 485.

21. J.F. Bernheisel, "Mid-Shakedown Evaluation of a Demonstration Resource Recovery Facility," in Municipal Solid Waste: Resource Recovery, Proc. Fifth Annual Research Symposium, U.S. EPA, MERL, EPA-6001-9-79-023b, August, 1979, p. 60.

22. J.F. Bernheisel, P.M. Bagelman, and W.S. Parker, "Trommel Processing of Municipal Solid Waste Prior to Shredding," in Proc. Sixth Mineral Waste Utilization Symposium, Eugen Aleshin ed., U.S. Bureau of Mines and IIT Research Institute, May 2-3, 1978, p. 255.

23. National Center for Resource Recovery, Inc., Trommel Initial Operating Report, Recovery 1, NCRR Technical Report TR-78-3, Washington, D.C., 1978.

24. American Iron and Steel Institute, Summary Report of Solid Waste Processing Facilities, AISI Washington, D.C., July, 1979.

25. Harvey Alter, S.L. Natof, K.L. Woodruff, and R.D. Hagen, The Recovery of Magnetic Metals from Municipal Solid Waste, NCRR, Inc., Washington, D.C., Nov. 1977.

26. G.M. Savage, L.F. Diaz and G.J. Trezek, "Performance Characterization of Air Classifiers in Resource Recovery Processing," in Proc. 1980 National Waste Processing Conf., ASME, May 11-14, 1980, p. 339.

27. J.G. Abert, Air Classification: A Vital Process, in NCRR Bulletin, Vol. 8, No. 1, NCRR, Washington, D.C., Winter, 1978, p. 101.

28. H.D. Funk and S.H. Russell, Energy and Materials Recovery System, Ames, Iowa, in Proc. Fifth Mineral Waste Utilization Symposium, Eugene Aleshin ed., U.S. Bureau of Mines and IIT Research Institute, April 13-14, 1976, p. 133.

29. G.F. Bourcier and K.H. Dale, "Aluminum Scrap Recovered From Full-Scale Municipal Refuse Processing Systems," in Proc. Sixth Mineral Waste Utilization Symposium, Eugene Aleshin ed., U.S. Bureau of Mines and IIT Research Institute, May 2-3, 1978, p. 179.

30. National Center for Resource Recovery, <u>Aluminum Recovery from MSW Using an Eddy Current Separator</u>, NCRR Technical Report, TR-80-8, Washington, D.C., June, 1980.

31. J.P. Cummings, Glass and Non-ferrous Metal Recovery Subsystem at Franklin, Ohio--Final Report, in <u>Proc. Fifth Mineral Waste Utilization Symposium</u>, Eugene Aleshin ed., U.S. Bureau of Mines and IIT Research Institute, April 13-14, 1976, p. 175.

32. J.J. Henn, <u>Updated Cost Evaluation of a Metal and Mineral Recovery Process for Treating Municipal Incinerator Residues</u>, BuMines, IC 8691, Washington, D.C., 1975.

33. E.A. Glysson, "Incineration Residue Separation--Recovery of Ferrous Fraction," in <u>Proc. 1976 Conf.--Present Status and Research Needs in Energy Recovery from Wastes</u>, Richard Matula Ed., ASME, New York, 1977, p. 229.

34. J.H. Heginbotham, Recovery of Glass from Urban Refuse by Froth Flotation, in <u>Proc. Sixth Mineral Waste Utilization Symposium</u>, Eugene Aleshin ed., U.S. Bureau of Mines and IIT Research Institute, May 2-3, 1978, p. 231.

35. J. Arnold, <u>Recovering Glass by Froth Flotation--Initial Operating Report</u>, NCRR Technical Report TR-77-8, Washington, D.C., Oct. 1978.

36. J. Arnold, <u>A Sizing Circuit for Preparation of Feedstock for Glass Recovery</u>, NCRR Technical Report TR-77-7, Washington, D.C., Oct. 1978.

37. J. Arnold, <u>The Preparation of Glass Feedstock and Non-ferrous Metals by Jigging</u>, NCRR Technical Report TR-80-7, Washington, D.C., June 1980.

38. National Center for Resource Recovery, Inc., <u>Preparation of Glass and Aluminum Concentrates using a Double-Deck Vibrating Screen</u>, NCRR Technical Report TR-80-7, Washington, D.C., June 1980.

39. D.S. Airan and J.H. Bell, "Resource Recovery through Composting--A Sleeping Giant," in <u>Proc. 1980 National Waste Processing Conf.</u>, ASME, May 11-14, 1980, p. 121.

40. C.G. Golveke, "Biological Processing: Composting and Hydrolysis," in <u>Handbook of Solid Waste Management</u>, D.G. Wilson ed., Van Nostrand Reinholt, New York, 1977, p. 197.

41. P. Gheresus, S.K. Adams, and J.C. Even, "Resource Recovery from Municipal Solid Waste: The Ames System Experience with Economics and Operation," in <u>Proc. 1980 National Waste Processing Conf., ASME</u>, May 11-14, 1980, p. 475.

42. Floyd Hasselriis, "The Greater Bridgeport, Connecticut Waste-To-Power System," in <u>Proc. 1980 National Waste Processing Conf., ASME</u>, May 11-14, 1980, p. 435.

43. C.R. Wiley and M.Bassin, "The Maryland Environmental Service/Baltimore County Resource Recovery Facility," in Proc. Sixth Mineral Waste Utilization Symposium, Eugene Aleshin ed., U.S. Bureau of Mines and IIT Research Institute, May 2-3, 1978, p. 281.

44. H. Alter and J. Arnold, "Preparation of Refuse-Derived Fuel on a Pilot Scale," in Proc. Sixth Mineral Waste Utilization Symposium, Eugene Aleshin ed., U.S. Bureau of Mines and IIT Research Institute, May 2-3, 1978, p. 171.

45. Carlton C. Wiles, "The Production and Use of Densified Refuse Derived Fuel," in Municipal Solid Waste: Resource Recovery, Proc. Fifth Annual Research Symposium, U.S. EPA, MERL, EPA-600/9-79-023b, August, 1979, p. 274.

46. National Center for Resource Recovery, Resource Recovery Activities, in NCRR Bulletin, Vol. 10, No. 1, March 1980, p. 17.

47. M. Dvirka and W.M. Harrington, Jr., "Update on Baltimore Pyrolysis Demonstration," in Proc. 1980 National Waste Processing Conf., ASME, May 11-14, 1980, p. 543.

48. J.P. Diebold and G.D. Smith, "Thermochemcial Conversion of Biomass to Gasoline," in Proc. EPA Conf. on Waste-to-Energy Technology Update 1980--Preprints, U.S. EPA, IERL, April 15-16, 1980, p. 39.

49. M.J. Blanchet, "Start-Up and Operation of the Landfill Gas Treatment Plant at Mountain View," in Municipal Solid Waste: Research Recovery Proc. Fifth Annual Research Symposium, U.S. EPA, MERL, EPA-600/9-79-023b, August, 1979 p. 345.

50. U.S. Environmental Protection Agency, Recovery of Landfill Gas at Mountain View/Engineering Site, EPA/530/SW-587d, Washington, D.C., May 1977.

51. L.F. Diaz, et al., "Biogasification of Municipal Solid Wastes," in Proc. 1980 National Waste Processing Conf., ASME, May 11-14, 1980, p. 403.

52. D.K. Walter and C. Rines, "Refuse Conversion to Methane (RefCOM)--A Proof-of-Concept Anaerobic Digestion Facility," in Proc. 1980 National Waste Processing Conf., ASME, May 11-14, 1980, p. 85.

53. Walter Brenner, et al., "Development of Continuous Acid Hydrolysis Process for the Utilization of Waste Cellulose," in Municipal Solid Waste: Resource Recovery, Proc. Fifth Annual Research Symposium, U.S. EPA, MERL, EPA-600/9-79-023b, August, 1979, p. 99.

54. V.R. Srinivasan, "Production of Methane from Acid Hydrosylates of Cellulose Wastes," in Municipal Solid Waste: Resource Recovery, Proc. Fifth Annual Research Symposium, U.S. EPA, MERL, EPA-600/9-79-023b, August, 1979, p. 114.

55. R.E. Frounfelker and N.J. Kleinhenz, "A Technical Evaluation of Modular Incinerators with Heat Recovery," in Proc. 1980 National Waste Processing Conf., ASME, May 11-14, 1980, p. 73.

56. M. Denchik, et al., "Akron Recycle Energy System is on the Line," in Proc. 1980 National Waste Processing Conf., ASME, May 11-14, 1980, p. 463.

57. W.C. Huang and D.L. Nelson, "Duluth Co-Disposal Facility," in Proc. 1980 National Waste Processing Conf., ASME, May 11-14, 1980, p. 551.

58. G.R. Fryling, Combustion Engineering, Revised Ed., Combustion Engineering, Inc., New York, 1966, p. 18-2.

59. R.E. Schwegler, and H.L. Hickman Jr., Report on Waste To Energy Projects in North America, in GRCDA Reports, Vol. 1, No. 1, GRCDA, Washington, DC, February 1981.

CHAPTER 5

COSTS

I. INTRODUCTION

The purpose of this chapter is to explain the process of estimating the capital and annual costs of the facilities needed in each of the selected alternatives. The required facilities include all of the land, buildings, and equipment needed for transfer stations, landfills and resource recovery plants. The costs for these facilities are used in conjunction with other previously gathered data in the following ways:

1. The transfer station costs are used along with the unit haul costs and transportation network data to optimize transfer station locations and to develop a system transport cost for each alternative.

2. The landfill costs form the basis for the system cost of the landfilling alternative, as well as the cost of disposing nonprocessable wastes, processing residue, and boiler (or incinerator) ash.

3. The resource recovery facility costs form the basis for the system costs of the resource recovery alternatives.

Since these facility costs are usually the major portion of the system costs (system costs include transport costs and disposal costs as well as facility costs), it is important that the investigator ensure that all of the applicable elements are included. It is not the purpose of the chapter to present the methodology for preparing a detailed cost estimate of the type necessary to be the basis for financing decisions. Detailed cost estimating is a specialized engineering field which requires the services of a professional engineer/estimator. The methodology presented in the following sections is, rather, for estimating costs on a level of detail appropriate to a preliminary feasibility analysis. At this level of detail, it is more important to include all of the major elements than to accurately list the cost for each "nut and bolt." The "big ticket" items are most important at this level of detail.

The following sections examine methods for estimating present-day costs in the general categories of Capital Costs, Annual Costs, Financing Costs, and Debt Service. The final section continues the River City example for numerical illustration.

II. CAPITAL COST ELEMENTS

Capital costs (sometimes called "first costs") include all expenditures related to the facility in question during the design, construction, and start-up of the facility. In order to accurately compare the capital costs of alternative systems, the following elements must be included for each: site development, buildings, equipment, rolling stock, energy distribution, contingency, engineering costs, legal costs, administrative costs, and construction insurance. This approach is especially applicable if the investigator wishes to compare the capital costs given by certain resource recovery "system" suppliers to other system alternatives. For example, many of these suppliers quote the capital costs for the equipment and buildings only. To accurately compare such a quotation to the alternatives, the investigator must add the costs of land, site development, rolling stock, energy distribution, engineering, legal, administration, and any financing costs which may be incurred. Without all of these elements, the capital cost estimate provided will be unrealistically low.

The first step in making a realistic capital cost estimate is to develop a preliminary site layout for the facility in question. A specific parcel of land for the facility need not be selected for this layout since it is only preliminary. For a landfill, the investigator should sketch the approximate size of the required perimeter areas (buffer zones), any roads, berms, or other site improvements which may be necessary, and scales, scale house, maintenance shed, or any other required buildings. For a transfer station, the drawing should include access roads, scales, the approximate size required for the receiving area, the location of major equipment (compactors, etc.), and the approximate size of the building. The drawing for the resource recovery facilities should include many of the same elements plus sketches of the processing lines and boilers with appropriate conveyors shown, RDF storage bins, and energy distribution equipment.

Through the exercise of making these preliminary site layout drawings the investigator will ensure that all the elements are included. Also, the drawings can be used for preliminary building sizes, for selecting an initial equipment list, and for calculating a preliminary site development cost. Following the preparation of these drawings, the capital costs can be calculated for each element. The following sections give the items which must be included along with guidelines for calculation.

A. Site Development

Site development includes items such as land purchase, earthwork, paving and entrance roads, site lighting, fencing and gates, landscaping, and utility connections. For landfills, additional items may include constructing surface runoff control structures (trenches, holding ponds), ground water monitoring wells, and gas collection wells.

Published cost estimating guides (1-4), which are updated annually, can be of assistance in determining current costs for these items. In addition, Reference 5 should be consulted for site development costs related specifically to landfills. Although the exact magnitude of these costs depend on the individual characteristics of the sites in question, site development costs exclusive of land purchase generally range between $3,000 to $4,000 per acre for landfills, and $30,000 to $60,000 per acre for transfer stations and resource recovery facilities (1981 costs). Unusual soil conditions and severe topographical variations could result in costs considerably higher than these ranges. The River City example at the end of this chapter illustrates typical ranges of site development costs for different facilities.

B. Buildings

Buildings are normally required to enclose the scale house, and equipment maintenance areas at landfills. At transfer stations and resource recovery facilities, building enclosures may include the receiving area, processing area, offices, locker rooms, maintenance shop, control room and boiler area. In mild climates, boiler enclosures and enclosures for truck manuvering space in the receiving area may be eliminated; however, enclosures for waste receiving areas and processing equipment are required to control dust, odor, and noise.

Published cost estimating guides (1-4,6) are helpful in determining building costs. Building costs are, of course, a function of the particular facility being analyzed. However, cost guidelines for the type of buildings required under average conditions can be found in Table 5.1. These costs are representative of the high-quality buildings necessary for a facility which is to be in operation for 20 to 25 years. These cost ranges are for general reference only, and should be tailored to the particular process under consideration. See the River City example for illustration of building costs for different facilities.

C. Equipment

A list of the equipment required for each facility should first be made. These equipment items include all of the stationary mechanical and

TABLE 5.1

1980 Building Cost Ranges

Building Type	1980 Capital Cost Range[a]
Transfer Stations (all areas)	$30 - $35 per sq. ft.
Tipping Floor and Truck Maneuvering Area	$35 - $40 per sq. ft.
Raw Waste Pit Storage[b]	$4.00 - $4.50 per cu. ft.
Processing Area	$45 - $60 per sq. ft.
Office, Control, Room, Shop, Locker Room, Scale house	$45 - $50 per sq. ft.
Boiler Enclosure	$3.00 - $3.50 per cu. ft.

[a]   Includes building enclosure, foundations, building electrical installation and building mechanical installation (excludes process electrical and mechanical installation).
[b]   Reinforced concrete pit, 30-foot depth.

electrical equipment required to put the facility into operation. A landfill will normally only require a scale since most other equipment is rolling stock. A transfer station may require scales, a stationary compactor, or a hydraulic clam-shell leveling arm. A resource recovery facility may require a wide variety of equipment including; shredders, trommel screens, magnetic separators, conveyors, boilers, waterwall incinerators, and turbine generators.

After making the equipment list for each facility, the investigator should consult certain published cost estimating guides (4) and equipment manufacturers for current budgetary costs. Certain resource recovery "system" suppliers will provide estimates of all the equipment needed in the facility; however, the supplier's list of equipment should be checked against the investigator's list to be certain that all items are included. The River City example gives typical equipment costs for different facilities.

D. Rolling Stock

Rolling stock includes earthmoving equipment and all vehicles except collection vehicles. A landfill operation may require scrapers for cover material, dozers and landfill compactors for refuse placement, water trucks for dust control, and pick-up trucks for general utility. The required number and type of vehicles and earthmoving equipment is a matter of landfill design and operation planning. The reference materials previously cited in Chapter 4 should be consulted for the preparation of a list of rolling stock for landfills.

A transfer station may require front-end loaders to move waste from a tipping floor, a sweeper for clean-up, general utility pick-up trucks, and transfer tractors and trailers. Note that although depreciation on the transfer tractor-trailers is included in the unit haul cost, the initial purchase cost must be included in the transfer station capital cost. The number of transfer tractors and trailers required depends on the size of the transfer station, its distance from the disposal site, and its operating schedule.

A resource recovery facility may require front-end loaders for tipping floor operation (if used), sweepers for clean-up, general utility pick-up trucks, dump trucks for residue and/or ash hauling, and transfer tractor-trailers for RDF hauling to a remote buyer. The number and type of vehicles required depends, of course, on the type of facility and its capacity.

The investigator should first list the rolling stock required for each facility, then contact a local equipment distributor for current budgetary cost estimates. The River City example illustrates cost estimates for rolling stock required for different facilities.

E. Energy Distribution

The capital costs for distributing recovered energy can be substantial. The investigator must, therefore, be certain that these costs are included in the capital cost estimates for resource recovery facilities. Any costs for modifying existing equipment at the energy market site (boilers, turbines, burners, etc.) must also be included.

For a facility producing RDF (refuse derived fuel--shredded particles), the costs for conveyors or truck transport of the RDF to the boilers of the energy market should be included. Generally, if the energy market is located more than 2,000 to 3,000 feet away from the processing plant, truck transport should be used. Truck transport requires a receiving building, surge storage, and a conveying system at the energy market site. Manufacturers of mechanical or pneumatic conveyors should be contacted to obtain pricing information on these systems if the transport distance is shorter than 2,000 to 3,000 feet. A gas production facility requires a pipeline to the energy buyer's boilers. Costs for underground or overhead pipelines can be obtained from pipeline manufacturers and installation costs can be calculated from published cost estimating references (1,2).

A steam production facility requires either an overhead or underground pipeline system. If condensate is to be returned, a return line must also be incorporated. There are several commercial manufacturers of steam/condensate pipeline systems. These manufacturers can be contacted for pricing information. Installation costs are available in published cost estimating guides (1-4).

An electric power or co-generation facility requires an electric power connection to a utility substation. The utility purchasing the power should be contacted regarding the location of the substation connection and the cost of constructing the required power feed line from the facility.

F.   Other Costs

There are a variety of other capital costs which must be included for a complete cost estimate. One of the most important is mechanical and electrical installation costs. These are the costs normally incurred by a general contractor for setting the equipment in place, making the required mechanical connections to other equipment, making piping connections, and making electrical connections. Since installation costs are a function of the amount of equipment to be installed, a percentage of the equipment cost can be used for estimation purposes. For the type of equipment required in a transfer station or resource recovery plant, a factor of 50 percent to 55 percent of the total uninstalled equipment cost (minus boilers, turbines, and other field-erected equipment) can be used with satisfactory accuracy in an initial feasibility analysis.

A construction contingency factor must also be included in the capital cost estimate. This factor is required to allow for change orders, unforseen construction delays, and other events which may cause the facility construction cost to be higher than expected. For an initial feasibility analysis, a factor of between 15 percent and 25 percent of the subtotal of all site development, buildings, equipment (installed), rolling stock, and energy distribution costs should be sufficient.

The costs for engineering, legal, and administrative services prior to and during construction should also be included in the capital cost estimate. A factor of 10 percent of all costs (including the contingency) is sufficient for the level of detail required in this type of analysis. Note that this cost does not include the costs which may be incurred in site selection and approval, and in obtaining major environmental permits.

A start-up reserve fund should also be capitalized to provide operating funds during the start-up period before full revenue payments begin. Experience with currently operating facilities indicates that 6 to

12 months is generally required for start-up activities. Operation and maintenance costs for this period should be included in the capital cost estimate.

III. ANNUAL COST ELEMENTS

Annual costs are the costs of operating and maintaining a facility after it has been constructed and started up. Although in practice these costs are incurred continuously, an annual summary is generally sufficient for estimating purposes. The elements which should be included in a complete annual cost estimate for a facility are given in the following paragraphs.

A. Maintenance

Maintenance items include site maintenance (mowing grass, snow removal, etc.) building maintenance (painting, minor repairs, cleaning, etc.), equipment maintenance (grease, oil, and other materials for the repair of conveyors, shredders, trommels, etc.), and rolling stock maintenance (gas, oil, tires, other repair materials, etc.).

For a landfilling operation, there is very little site and building maintenance required; however, an annual allowance for installation of gas collection wells as the fill progresses should be included if such wells are part of the design. An annual cost for quarterly ground water sampling and analysis should also be included if monitoring wells are included. (See Reference No. 7 for cost factors.) Most of the landfill maintenance cost will be in rolling stock operation and repair. Equipment suppliers can provide estimates of operating and maintenance costs for vehicles and earthmoving equipment.

For a transfer station or resource recovery plant, site maintenance will vary depending on climate and other conditions. For estimating purposes, however, between $750 and $1000 per acre per year (1980 dollars) is generally sufficient. Building maintenance costs are also variable depending on the type of building, its size, and the maintenance schedule. Generally between $0.75 and $1.00 per square foot of floor area per year (1980 dollars) is sufficient for estimating purposes. The investigator could also check the annual building maintenance costs of similar local municipal buildings. Equipment maintenance costs for a transfer station or resource recovery plant is highly variable and is a function of equipment reliability, operation schedule, refuse composition and the quality of the maintenance staff. For preliminary annual cost estimating purposes, however, a factor of 5 percent per year of the delivered, uninstalled equipment cost can be used for well-maintained refuse handling and processing equipment under normal operating circumstances. For boiler and power generation equipment, a factor of 2 percent to 3 percent of the delivered, uninstalled capital cost per year can be used with satisfactory accuracy. The one exception to the use of these factors is the raw refuse shredder. It is a high-maintenance equipment item for which maintenance costs should be calculated separately. For a hammermill type shredder which requires hammer re-tipping and/or replacement periodically, a factor of between $0.50 and $1.00 per ton processed (1980 dollars) is reasonably accurate for estimating purposes.

Rolling stock maintenance for all vehicles in transfer stations and resource recovery plants should also be included. Note that because the operating and maintenance cost of the transfer tractor/trailers is included in the unit haul cost, these rolling stock maintenance costs should not be included in the transfer station annual cost estimate. For the same reason, the maintenance costs for ash and residue haul vehicles should not be included in the resource recovery facility costs. Other rolling stock maintenance costs (front-end loaders, pick-up trucks, sweepers, etc.) should, however be included in the annual cost estimate for these facilities. Cost factors can be obtained from equipment manufacturers. The River City example illustrates the magnitude of certain annual maintenance costs for different facilities.

B. Labor

Labor costs include the wages, fringe benefits, and related administrative costs of the manpower to operate and maintain the required facilities.   Labor cost is usually one of the largest single annual cost line items, so careful consideration is warranted.   Local labor rates for the study area should be used for this calculation. Labor unions can be contacted for wage rates in the labor categories required.   If municipal employees are to be utilized, the wage rate schedule can be obtained from the municipality.   The investigator should also determine the escalation clauses included in present labor union contracts or municipal wage schedules.   Fringe benefits and administrative costs are highly variable, but generally range from 20 percent to 40 percent of the base wage.

    The first step in determining the labor cost for a facility is to develop a labor schedule which lists all of the job categories required, the number of employees required in each category per shift, the number of shifts required for each job category, and the resulting total required employees in each job category.   The numbers and types of employees needed at a facility depends, of course, on the type of facility, its size, local union contract provisions, and other factors.

    The River City example in Section V of this Chapter illustrates the labor schedule required in certain situations.   Although it is not possible to describe the labor requirements of all variations of facilities, the general considerations covered in the following paragraphs should provide guidance.

    1. Landfills

In general, a landfill will require a superintendent, foremen, scale operators, dozer operators, mechanics, and laborers.   For large landfill operations, traffic directors, scraper operators, compactor vehicle operators, and clerks may also be required.   Many landfills receive

waste six days per week, eight-to-ten hours per day. See the River
City example for typical landfill labor schedules.

2. Transfer Stations

A transfer station requires both operators and transfer vehicle drivers.
Operating and maintenance personnel may include foremen, scale
operators, equipment operators (for compactors, front-end loaders,
leveling arms), and laborers. For small transfer stations, one person
can handle more than one of these functions. The number of transfer
vehicle drivers required depends not only on the design and capacity of
the transfer station, but also on the distance between the transfer
station and the final disposal site (or resource recovery facility). The
number of drivers required will parallel the required number of transfer
tractors. Since neither the transfer station locations nor the locations
of the other facilities have been finalized at this point in the analysis,
it is impossible to define exactly the number of required drivers. By
examining the map locations of the potential facility sites, however, the
investigator can choose an approximate average haul distance for use in
selecting transfer trailer driver requirements. The River City example
(Table 5.3) illustrates the approximate number of drivers required for
three sizes of transfer stations assuming a 50-mile round trip haul
distance.

3. Resource Recovery Facilities

The manpower schedules required for resource recovery facilities are as
varied as the range of sizes and technologies available. In general,
however, the personnel requirements of a resource recovery facility may
be: a superintendent (who may share other public works duties if
municipally operated), a foreman for each shift, scale operators for
shifts during receiving hours, traffic directors, front-end loader
operators, crane operators, processing line operators, mechanics and
electricians for maintenance, laborers, truck drivers for residue/ash
hauling, and steno/clerks. For facilities including boiler and power
generation equipment, an additional foreman may be required, along

with boiler operators, and turbine operators. If the facility is to be operated on a 24-hour per day basis, a fourth shift, or "swing" shift should be included.

C. Utilities

The annual costs of utilities should be part of the annual cost estimate. These would include natural gas or heating oil for building heat, domestic water and sewer, telephone, and electric power. Electric power will be the major utility cost for a resource recovery plant. The costs of other utilities are mostly a function of building type and size and can be estimated from data supplied by local utilities or by examining usage data for similar buildings in the study area.

Electric power requirements for resource recovery facility equipment should be calculated separately. Since solid waste processing equipment and boiler equipment require large amounts of power for operation, electric power costs can be substantial. Power requirements vary widely depending on the amount and type of processing and boiler equipment used. Processing equipment power requirements can range from 20 to 60 KWH per ton processed. Boiler equipment power requirements can range from 5 to 10 KWH per thousand pounds of steam produced for waterwall incinerators and stoker-fired travelling grate boilers. Fluid bed reactors have higher power requirements. Electric power costs can be included as an annual utility cost line item. For facilities which produce electric power, however, a common practice is to provide internal power, and sell a net amount. Either method will properly account for this major cost item.

D. Residue/Ash Hauling and Disposal

The costs of hauling and disposing the processing residue and/or ash from a resource recovery facility must be included in the annual costs for these facilities. Residue from waste processing will be produced at a rate of up to 25 percent of the incoming raw refuse depending on the refuse composition and the type and degree of processing utilized. Ash production will range from 8 percent to 35 percent of the incoming raw waste depending on waste composition, the degree of pre-processing, and the combustion (or pyrolysis) equipment used (see Chapter 4, Table 4.4). Using the appropriate factors for the resource recovery facility in question, quantity estimates can be made.

Costs for disposal of these materials can be calculated by using the current per-ton landfilling costs in the study area, or alternatively can be covered in the annual operating costs of a support landfill for residue, ash, and non-processables. Costs for hauling residue and/or ash from the resource recovery facility site(s) is calculated from the unit haul cost developed in Chapter 2, the residue and ash quantity estimates, and the haul distance. These calculations must, however, await the systems analysis (Chapter 6) in order to derive haul distances because facility locations have not as yet been selected.

E.   Other

Other annual cost elements include; contingency, and insurance. A contingency should be included in the annual cost estimate of a facility to account for unforseen equipment breakdowns, labor strikes, unusually high maintenance costs, and other factors which may increase annual costs. For an initial feasibility analysis, a factor of 10 percent of the sum of all maintenance, labor, utilities, and residue/ash haul and disposal costs should be sufficient.

The annual cost for operating insurance is highly variable and depends on location, type of facility, facility size, degree of coverage, magnitude of deductibles and other factors. In general, the type of coverage which should be considered includes: workmen's compensation, comprehensive general liability, automobile liability (for rolling stock), property damage (all risk), and business interruption insurance. The projected annual cost for such coverage in the Pinellas County, Florida, project for the start-up year (1984) is $168,000 (8) which is about 0.3

percent of the unescalated construction cost. This percentage may not, however, be applicable to other projects because insurance costs are highly dependent on the magnitude of the deductibles. An industrial insurance underwriter should be contacted for an estimate.

IV. FINANCING COSTS AND DEBT SERVICE

This section is concerned with two separate but related components of the capital and annual costs of a facility: financing costs (capital costs) and debt service (annual cost). The sum of all of the capital cost elements previously presented will provide enough funds to construct and start-up the facility. There are, however, other capital costs which will be incurred to obtain debt financing for these funds. These are "up front" costs which should be distinguished from annual interest costs. The magnitude of these financing costs depends on the financing method chosen. Annual debt service is related to financing costs because the financing costs are a component of total capital cost. The calculation of financing costs and debt service is a specialized area of expertise requiring the services of qualified professionals.

In order to calculate financing costs and the related debt service, capital and annual costs must first be adjusted for inflation. To account for capital cost inflation, it is necessary to know when the construction dollars will be spent. Also, annual costs must be adjusted for inflation to accurately reflect the costs of operation in the first operating year. In order to make these adjustments, the investigator must develop an implementation schedule. Implementation planning is discussed in Chapter 7, however a detailed plan is not required at this point in the analysis. The implementation schedule should allow time for design development and the securing of construction and environmental permits. A time period for developing the financing package and securing the funds should also be allowed. The actual construction period and start-up period should also be shown in the schedule. Depending on the type and size of the facility, and the complexity of the financing package and required institutional arrangements, the total period could range from two to six years. The investigator should consult Chapter 7 of this book for more detail about implementation planning.

For large facilities, the construction period may be two to three years in length. Capital costs can be escalated from present dollars to the mid-point of the construction period for simplicity. For a six-year total implementation schedule with a three-year construction period, for example, the escalation period would be four and one-half years. Estimating the magnitude of construction cost inflation over a four - to five - year projection period is certainly speculative; however, historical information is available in certain published construction cost indicies (9). Rates of 10 percent to 12 percent per year (compounded) have been used.

Escalating operation and maintenance costs to the point in time at which the facility is in full-scale operation is also a speculative exercise. The investigator should examine the cost-of-living adjustment clauses and other escalation factors built in to existing union labor contracts to estimate labor cost escalation rates. Local utilities can be contacted to develop estimates of energy cost increases. Other maintenance costs increases will normally lag behind labor and energy cost escalation rates by one to three percent per year.

With capital costs escalated to construction-year dollars and annual operation and maintenance costs escalated to operation-year dollars, the financing costs and debt service can be calculated. Although a detailed analysis requires the services of a qualified professional, the following paragraphs describe in conceptual terms certain available financing methods and the relative magnitude of the financing costs and debt service requirements of each (10,11). In each of the following cases the interest income to investors is tax-exempt, making the interest rates lower than taxable investments. Taxable debt financing can also be used if other factors outweigh the higher interest rates.

A. General Obligation Bonds

General obligation financing by a municipality, county, authority, or state government is usually the lowest cost debt instrument in terms of interest rate available (assuming a favorable credit rating for the issuing entity). The issuance of these bonds must be supported by referendum because the "full faith and credit" of the issuing entity is pledged to the repayment of the principal and interest to the bond holders. The bond issue size is also usually limited by the constitutional debt ceiling established for the issuing entity. The Ames, Iowa, Solid Waste Recovery System utilized this financing method as the major source of construction funds.

The advantages of this financing method are its relatively low interest rate and hence debt service, and the relative simplicity of the required institutional arrangements. Disadvantages include high risk exposure to the issuing entity and its taxpayers, the time lag in passing a referendum, the elimination of facility tax payments because of public ownership, and the unavailability of federal tax benefits available to private owners of resource recovery facilities (investment tax credits and accelerated depreciation). Also, because of the debt ceiling restriction, the funds used for the facility reduces the amount of dollars which can be used by the issuing entity for other projects such as streets, sewers, and wastewater treatment plants.

The capitalized financing costs required for a general obligation debt instrument will, in general, include a fund to cover interest accrued on the borrowed funds during the construction period (IDC), bond issuance expenses (or underwriter's "spread") minus interest earned during the construction period on the investment of the construction funds and the capitalized IDC fund. This interest income during the construction period is possible because the construction funds and IDC funds are not paid out in a lump sum. Since these funds are paid out gradually over the construction period, part of the funds are available during a portion of the construction period for investment on a short-term basis. Although a function of a variety of factors specific to each bond issue, the net financing costs of a general obligation bond issue generally will be between 2 percent and 10 percent of the total construction cost (including contingency, engineering, legal, administration, construction insurance, and start-up fund).

The annual debt service which corresponds to this financing method is calculated by multiplying the total capital cost requirement (escalated construction cost plus all financing costs) by the capital recovery factor for the bond interest rate and the financing period of the financing package. Mathematically,

$$A = TCC \ \frac{i \ (1+1)^n}{(1+i)^{n-1}} \tag{1}$$

where:

A   = Annual debt service in dollars per year
TCC = Total capital cost in dollars
i   = Bond interest rate in percent per year
n   = Financing period in years

Tables for capital recovery factors are available in numerous references (12).

B.   Revenue Bonds

There are a number of kinds of revenue bonds which can be used to finance a resource recovery facility. They have in common a financial structure which utilizes project revenues (tipping fees, recovered material and energy sales, etc.) as the primary source of bond repayment. The following paragraphs describe some of the available methods.

1.   Non-Profit Public Corporation

This form of financing utilizes a non-profit corporation backed by a city, county, authority, or state government which issues revenue bonds to construct and own the facility. Since the basic structure is a project revenue bond, this financing mechanism is only applicable to a facility which is essentially self-supporting. Financing is possible with this method for a landfill or transfer station (in addition to a resource recovery facility) as long as the tipping fees are structured to cover all costs.

The backing comes from the pledge of the supporting entity's excise taxing power as a last line of repyament in case project revenues cannot cover costs. This type of financing, of course, applies only to entities which have excise taxing power. The facility would be owned by the non-profit corporation which can either lease it to the operating entity or sell it over time.

The advantages of the non-profit corporation financing approach include the fact that it has been used successfully to finance other types of public projects (including housing rehabilitation, pollution control), its avoidance of the constraint of constitutional debt limitations, and the elimination of the need for a referendum. Also, the interest rate and debt service should be lower than other types of revenue bonds because of the excise tax backing. Interest rates are, however, likely to be slightly higher than general obligation interest rates. Disadvantages include the fact that the facility is not on the tax rolls, and federal tax benefits are not available to private owners of resource recovery facilities.

Capitalized financing costs for this financing mechanism generally include those previously discussed for the general obligation financing plus a bond reserve fund which is often equivalent to a year of principal and interest payments. Buyers of revenue bonds will often require such a reserve fund to guard against a catastrophic failure of equipment. Many times in order to successfully market the bonds, the project must show a revenue income which not only covers all operation and maintenance costs, but results in an excess of 15 percent to 20 percent of the annual debt service in addition to this capitalized reserve fund. The investment earnings on this bond reserve fund during construction can, however, be subtracted from the total bond issue size. Further, assuming that the bond reserve fund is not used, additional annual earnings from the investment of this fund can be subtracted from the annual operating costs of the facility. In general, the net capitalized financing costs of this mechanism range from 15 percent to

30 percent of the total construction cost. Annual debt service is calculated as previously described in equation (1).

2. Industrial Revenue Bond--Tax Exempt Entity

Similar to the non-profit corporation revenue bond, this approach utilizes a project revenue bond. The difference is that a city, county, authority, or state issues the bonds without the excise tax backing. These bonds, therefore, must offer a higher rate of interest to buyers in order to be marketable. Interest rates of 2 percent to 2.5 percent higher than the non-profit corporation bonds may be required. Again, with this mechanism the facility will not be on the tax rolls and federal tax benefits will not be available because of public ownership.

Capitalized financing costs for this mechanism are similar to those required for the non-profit corporation bonds. IDC and bond issuance costs will be higher because of the higher interest rates and the greater marketing difficulty. In general, these capitalized financing costs will range from 25 percent to 50 percent of the total construction costs. Annual debt service is calculated as previously described in equation 1.

3. Industrial Revenue Bond--Taxable Entity

This method differs from the other revenue bond issues previously discussed because the bonds are issued by a city, county, authority, or state government on behalf of a private corporation. The funds are then loaned to the private corporation for construction and operation of the facility. This method of financing was utilized in the Saugus, Massachusetts, resource recovery project. These bonds can be issued for the purpose of creating industrial development in a community or for building pollution control equipment. In the first case, the term, "Industrial Development Bond (IDB)" has been used, and in the second case, the term, "Pollution Control Revenue Bond (PCRB)" has been used for resource recovery projects.

Because the private corporation assumes the risks of construction completion, economical operation, and the repayment of the debt, potential bond buyers will scrutinize the credit-worthiness of the corporation when analyzing a bond purchase. It is important, therefore, that the private corporation chosen for such a project be as financially sound as possible.

The advantages of this financing mechanism include a shift of the majority of project risks away from the municipality, the payment of property and income taxes by the operators of the facility, and the availability of certain federal tax benefits to the private corporation if the facility is a resource recovery plant. Although many project risks are shifted to a private corporation, the corporation will expect to make a profit from operating the facility. This may make the system cost of solid waste disposal to the community higher than with other financing methods. The interest rate (and hence debt service) will be roughly equivalent to a revenue bond issue by a tax-exempt entity, assuming that a suitably credit-worthy corporation can be found to undertake the project.

Capitalized financing costs will generally include the elements previously described for other revenue bond issues. These financing costs will, however be higher because the interest earned on the construction fund, capitalized IDC fund, and the bond reserve fund during the construction period is taxable, which results in a lower net yield. Financing costs can be expected to be between 40 percent and 50 percent of the total construction cost for this financing mechanism. Annual debt service is calculated as previously described in equation 1.

C. Leveraged Lease

The leveraged lease financing concept has been used for many years in the financing of aircraft, rail cars, and other transportation equipment; warehousing; and other facilities. Such a concept in the financing of resource recovery facilities may produce a significantly lower net borrowing cost than revenue bond financing, although certain execution complexities are entailed.

The leveraged lease transaction can be structured like a bond financing execpt that an equity investor, or group of equity investors (lessor) would own and lease the facility to an operator which could be either a municipality or a private entity (lessee). These equity investors typically are large, credit-worthy taxable institutions (banks, insurance companies), or other private companies. The equity investors contribute part of the total required capital, with the remainder contributed through the issuance of tax-exempt bonds (either General Obligation or Revenue Bonds) by the affected municipality or through taxable debt from a private lender. The proportion typically is from 50 percent equity/50 percent debt, to 20 percent equity/80 percent debt. Lease payments by the lessee would cover debt service on the bonds (or loan) and would take into account the tax benefits which accrue to the equity investors.

The method is called "leveraged lease" because the equity investors can "purchase" the tax benefits for the entire project by contributing as little as 20 percent to 25 percent of the total cost according to Internal Revenue Service (IRS) rules. These benefits can then be passed along to the municipality through lower lease payments. Tax benefits currently available include accelerated depreciation (ACRS), a 10 percent investment tax credit (ITC), and the new energy tax credit (ETC). The availability of these tax benefits are subject to a number of IRS regulations and "tests." An IRS ruling may be required before the exact magnitude of these benefits can be defined. Institutional investors have indicated much interest in purchasing ownership interests in resource recovery facilities. Although the investments of such equity participants would be subordinate to the interest of the bondholders (or lender), the equity investors may be entitled to certain federal tax benefits, which is their primary interest. They are only secondarily interested in cash return, and as a result, the annual debt service may be substantially reduced since the investor's equity return is mostly generated by federal tax benefits and not by cash from project revenues. In addition, the equity investors would hold title to the facility at the end of the twenty to twenty-five-year operating period and thus would benefit from its residual value at that time.

The major advantage of the leveraged lease financing method is that it reduces the size of the bond issue which results in a proportionate decrease in annual debt service. The higher the percentage of equity contribution from investors, the greater the benefit to the municipality. The major disadvantage is the complex legal structure of the transaction. Capitalized financing costs will include components similar to the revenue bond with a taxable entity mechanism. Leveraged lease financing costs will, however, be lower in terms of percentage of construction cost due to the equity contribution which lowers the number of bonds to be sold (or other debt). Annual debt service will also be lower than 100 percent debt financing to the extent of the proportion of equity contribution. The proportion of equity contribution is determined in part by a final determination of which tax incentives are available to the investors. This determination is a function of the type of resource recovery facility and continuing legal interpretation by the IRS and the courts. A general discussion of the leveraged lease transaction can be found in the reference materials (13).

D. Municipal Lease Purchase

A relative newcomer to resource recovery facility financing is the Municipal Lease Purchase method. The lease purchase concept is structured such that an equity investor or group of equity investors contribute 100 percent of the construction cost and lease the facility to a municipality. The municipality is then obligated to make periodic payments of principal and interest in much the same manner as with a bond debt service. Current legal opinions indicate that the interest earned by the investors is exempt from state and federal taxes, and investors may be able to take advantage of the same federal tax benefits available under the leveraged lease transaction.

There are several firms in existence which offer the lease purchase arrangement. This financing package has been used by cities, states, counties, and special districts (such as hospital, fire, sewer) mainly for the purchase of equipment (computers, telephone systems, vehicles, and medical equipment), although some buildings have been financed (14). A $23.5 million port facility expansion in Portland, Oregon, has been financed with this method. Certificates of Participation are sold much like bonds to investors, or to brokers who sell to investors.

This financing arrangement was set up originally to bridge the gap between municipal capital investments small enough to be handled with annual appropriations and larger investments which required the sale of bonds. Interest rates will be higher than a general obligation bond for the same municipality because of higher investment risk. Up front financing costs will be similar to a general obligation bond issue.

V. RIVER CITY COSTS

The following is a presentation of the cost analysis of the facilities required for each of the alternatives selected for River City in Chapter 4. Transfer station costs are presented first followed by the costs for the landfill and resource recovery facilities required in each system alternative. In Chapter 6 of this book, these facility costs will be combined with transport costs and revenues (where applicable) to compute net system costs for each alternative.

The following general assumptions were made in the preparation of the River City facility costs:

1. The new landfill and resource recovery facilities were sized to handle the entire processable waste stream generated in the 20-year operating period between 1985 and 2005 (approximately 9 million tons). The primary facilities in each alternative system have design capacities of about 400,000 tons per year to approximately match quantity projections for 1985 (see Chapter 4).

2. All facilities were assumed to begin full-scale operation in 1985. Transfer stations and landfills may have shorter implementation schedules than resource recovery facilities, but were assumed to

have the same initial operation date for consistancy. The mid-point of construction for all transfer stations and resource recovery facilities was assumed to be 1983, the mid-point assumed for the landfill was 1984.

3. Calculations confirmed that the existing landfills have sufficient capacity for the 20-year disposal of non-processable wastes, processing residue and/or ash in all alternative systems. It is, therefore, not necessary to construct new landfill capacity to support any of the resource recovery alternatives, nor is additional landfill space for non-processable wastes required in the landfilling alternative.

4. Construction costs were estimated in 1980 dollars, then escalated to the mid-point of the construction period at 12 percent per year (compounded).

5. Operation and maintenance costs were also estimated in 1980 dollars, then escalated to the 1985 base operating year at an assumed average inflation rate of 10 percent per year (compounded).

6. General obligation financing at 8 percent for 20 years is assumed for the transfer stations and the landfill. Project revenue financing at 9 percent for 20 years is assumed for the resource recovery alternatives.

A. Transfer Station Costs

Tables 5.2 and 5.3 display the capital and annual cost estimates for the three sizes of transfer stations selected for analysis. Capital costs range from about $1.5 million for the single-compactor 1,250 TPW station, to about $2.8 million for the 3,758 TPW station. Note the

TABLE 5.2

River City Transfer Station Capital Costs

Item	Capital Costs ($ x 1000)		
	1,250 TPW[a]	2,500 TPW[b]	3,750 TPW[c]
1. Site Work	182	273	364
2. Buildings	153	234	294
3. Stationary Equipment	128	159	190
4. Mobile Equipment	62	62	62
5. Transfer Tractors	80	120	210
6. Transfer Trailers	120	150	210
7. Subtotal (items 1-6)	725	998	1,280
8. Contingency (15% of item 7)	109	150	192
9. 1980 Const. Cost (items 7 & 8)	834	1,148	1,472
10. 1983 Const. Cost[d]	1,171	1,613	2,068
11. Eng.,Legal, Admin. (10% of item 10)	117	161	207
12. Start-up reserve[e]	98	175	230
13. Subtotal (items 10-12)	1,386	1,949	2,505
14. Financing Costs (10% of item 13)[f]	139	195	251
15. Total Capital (items 13 & 14)	1,525	2,144	2,756

[a] Single-compactor design.
[b] Double-compactor design.
[c] Triple-compactor design.
[d] Escalation from 1980 to 1983 at 12% per year.
[e] Six months of operation and maintenance costs (see Table 5.3).
[f] Assuming general obligation bond financing.

TABLE 5.3

River City Transfer Station Annual Costs

Item	Annual Costs ($ x 1000)		
	1,250 TPW[a]	2,500 TPW[b]	3,750 TPW[c]
1. Site & Bldg. Maintenance	6	8	10
2. Mobile Equip. Fuel and Maintenance	18	18	18
3. Utilities	7	11	14
4. Labor[d]	80	160	220
5. Subtotal (items 1-4)	111	197	262
6. Contingency (10% of item 5)	11	20	26
7. 1980 Annual Cost (items 5 & 6)	122	217	288
8. 1985 Annual Cost[e]	196	349	464
9. Debt Service[f]	155	218	281
10. Total Annual (items 8 & 9)	351	567	745

[a] Single-compactor design, 2 operators, 2 drivers, 50-mile round trip transfer.
[b] Double-compactor design, 3 operators, 5 drivers, 50-mile round trip transfer.
[c] Triple-compactor design, 4 operators, 7 drivers, 50-mile round trip transfer.
[d] Average annual salary plus fringes and overhead is $20,000.
[e] Escalation from 1980 $ to 1985 $ at an average of 10% per year.
[f] 8% interest rate, 20-yr. period.

economy of scale evident in these construction costs. The largest station has three times the capacity of the smallest station, but less than double the capital cost. The 1985 annual costs (including debt service) range from about $350,000 for the smallest station to about $750,000 for the largest. Again, the economy of scale is evident.

B. Alternative 1--Landfill

The capital and annual cost estimates for a new, 640-acre sanitary landfill are displayed in Tables 5.4 and 5.5. These estimates assume a landfill designed and operated in accordance with the federal guidelines (see Chapter 4). The capital cost requirement is about $9 million, with a 1985 annual cost (including debt service) of about $3 million for a 7,700 TPW operation. Table 5.6 is the labor schedule assumed in the development of the annual costs. It is assumed that these cost estimates are equally applicable to either potential landfill location.

C. Alternative 2--Steam to RCI

Figure 5.1 is a preliminary site layout for the mass-burn steam plant required in this alternative. The major equipment items and approximate building sizes are shown. Tables 5.7 and 5.8 present the capital and annual cost estimates for this facility. About $95 million in capital cost will be required with a 1985 annual cost of about $16 million for a design capacity of 7,200 TPW. Table 5.9 summarizes labor schedule assumed for this facility.

D. Alternative 3--Co-generation

The preliminary site layout for this facility (Figure 5.2) is quite similar to Figure 5.1. A turbine room and cooling towers have been added to

TABLE 5.4

River City Alternative 1 - Landfill, Capital Costs
(7,700 TPW)

Item	Capital Costs ($ x 1000)
1. Land (640 Ac)[a]	1,920
2. Site Work:	
Boundary fence, roadways, groundwater monitoring wells, leachate collection system, leachate treatment system, scales, landscaping, and litter fences	760
3. Buildings:	
Maintenance building and equipment storage (100' x 40' x 35') and scale house	150
4. Rolling Stock:	
Compactor(1), dozers(2), scrapers(2), motor grader(1), water truck(1), and pick-up trucks (2)	1,340
5. Subtotal (items 1-4)	4,170
6. Contingency (15% of item 5)	626
7. 1980 Construction Cost (itme 5 & 6)	4,796
8. 1984 Construction Cost[b]	7,547
9. Eng., Legal, Admin. (10% of item 8)	755
10. Subtotal	8,302
11. Financing Costs (10% of item 10)[c]	830
12. Total Capital (items 10 & 11)	9,132

[a] 9 million tons, 800 lb/CY in-place density, 30 foot fill depth, 4:1 ratio, 15% for perimeter areas, buildings, and roads.
[b] Escalation from 1980 $ to 1984 $ at 12% per year.
[c] Assuming general obligation financing.

TABLE 5.5

River City Alternative 1 - Landfill, Annual Costs
(7,700 TPW)

Item	Annual Costs ($ x 1000)
1. Building and Entrance Maintenance	10
2. Labor[a]	420
3. Rolling Stock Operation and Maintenance	510
4. Final cover and landscaping[b]	50
5. Lining[c]	300
6. Leachate Treatment System Operation	10
7. Subtotal (items 1-6)	1,300
8. Contingency (10% of item 7)	130
9. 1980 O & M Costs (items 7 & 8)	1,430
10. 1985 O & M Costs[d]	2,303
11. Debt Service[e]	930
12. Total Annual Cost (items 10 & 11)	3,233

[a] 21 total employees, see Table 5.6.
[b] 32 acres per year.
[c] Plastic liner: 22 acres per year.
[d] Escalation from 1980 $ to 1985 $ at 10% per year.
[e] 8%, 20-year financing of Total Capital from Table 5.4.

TABLE 5.6

**River City Alternative 1 - Landfill, Labor Schedule
(7,200 TPW)**

Job Category	No. per Shift	No. of Shifts	Total Required
Superintendent	1	1	1
Forman	1	2	2
Scale Operator	1	2	2
Traffic Director	1	2	2
Compactor Operator	1	2	2
Dozer Operator	2	2	4
Scraper Operator	2	2	4
Mechanic/Oiler	1	1	1
Laborer	1	2	2
Steno/Clerk	1	1	1
Total			21

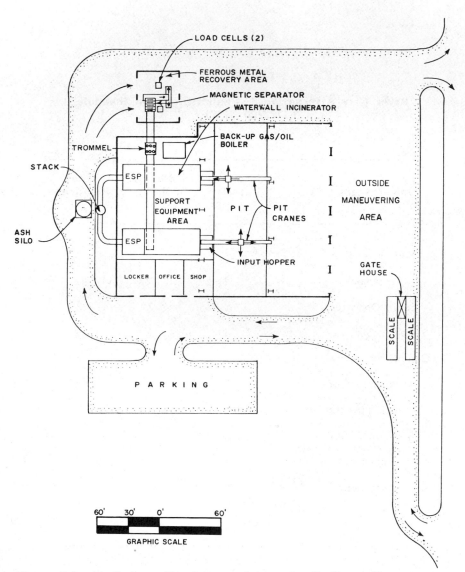

LOAD CELLS (2)

FERROUS METAL
RECOVERY AREA

MAGNETIC SEPARATOR

WATERWALL INCINERATOR

BACK-UP GAS/OIL
BOILER

TROMMEL

STACK

ESP

SUPPORT
EQUIPMENT
AREA

PIT   PIT
CRANES

OUTSIDE

MANEUVERING

AREA

ASH
SILO

ESP

INPUT HOPPER

GATE
HOUSE

LOCKER   OFFICE   SHOP

SCALE   SCALE

PARKING

60'   30'   0'   60'

GRAPHIC SCALE

Figure 5.1.   Preliminary Site Layout--Alternative 2, Steam Plant

TABLE 5.7

River City Alternative 2 - Steam to RCI, Capital Costs (7,200 TPW)

Item	Capital Costs ($ x 1000)
1. Site Work	500
2. Buildings: raw refuse pit, incinerator building, and receiving building	9,310
3. Mechanical Equipment: pit cranes(2), scales(2), dust collection system, installation	2,480
4. Steam Generation Equipment: two, 800 TPD (220,000 pph steam) waterwall incinerators (60 psig/sat.), pumps, fans, stacks, water treatment, other auxillary equipment, one 100,000 pph (steam) gas/oil package boiler, and installation	26,400
5. Rolling Stock: ash haul trucks, pick-up, and sweeper	290
6. Steam Line	500
7. Subtotal (items 1-6)	39,480
8. Contingency (15% of item 7)	5,922
9. 1980 Construction Cost (items 7 & 8)	45,402
10. 1983 Construction Cost[a]	63,786
11. Eng., Legal, Admin. (10% of item 10)	6,379
12. Start-up Reserve[b]	2,657
13. Subtotal (items 10-12)	72,822
14. Financing Costs (30% of item 13)[c]	21,847
15. Total Capital (items 13 & 14)	94,669

[a] Escalation from 1980 $ to 1983 $ at 12% per year.
[b] Six months of operation and maintenance costs, see Table 5.8.
[c] Assumes revenue bond issue by the City.

TABLE 5.8

River City Alternative 2 - Steam to RCI, Annual Costs (7,200 TPW)

Item	Annual Costs ($ x 1000)
1. Site and Bldg. Maintenance	50
2. Mechanical Equipment Maintenance	80
3. Steam Equipment Maintenance	530
4. Labor[a]	1,080
5. Electric Power	700
6. Other Utilities	20
7. Ash Haul and Disposal[b]	540
8. Subtotal (items 1-7)	3,000
9. Contingency (10% of item 8)	300
10. 1980 O & M Cost (items 8 & 9)	3,300
11. 1985 O & M Cost[c]	5,315
12. Debt Service[d]	10,372
13. Total Annual Cost (items 11 & 12)	15,687

[a] 54 employees at an average annual salary, overhead, and fringe benefits cost of $20,000, see Table 5.9.
[b] Assume 46 miles round trip, 30% of raw waste.
[c] Escalation from 1980 $ to 1985 $ at 10% per year.
[d] 9%, 20-year financing of Total Capital from Table 5.7.

TABLE 5.9

River City Alternative 2 - Steam to RCI, Labor Schedule (7,200 TPW)

Job Category	No. per Shift	No. of Shifts	Total Required
Superintendent	1	1	1
Foreman	1	4	4
Scale Operator	1	2	2
Traffic Director	1	2	2
Crane Operator	2	4	8
Boiler Operator	2	4	8
Ash Loader	2	4	8
Laborer	2	2	4
Mechanic	1	2	2
Electrician	1	2	2
Truck Driver	3	1	3
Relief Operators	2	4	8
Steno/Clerk	2	1	2
Total			54

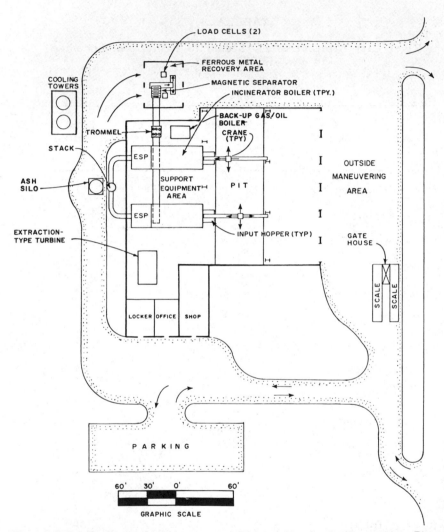

Figure 5.2.　Preliminary Site Layout--Alternative 3, Co-Generation Plant

the drawing. Tables 5.10 and 5.11 give the capital and annual cost estimates for this facility. Capital costs total almost $120 million, with a 1985 annual cost of about $18 million for a design capacity of 8,400 TPW. Table 5.12 is the labor schedule used for the annual cost calculation.

E. Alternative 4--Steam for District Heating System

Figure 5.3 is a preliminary layout of the RDF processing plant after modification of the existing transfer station. It was assumed that a portion of the tipping floor, the scales, and a portion of the roads would be re-used in the new facility. The major processing equipment and a portion of the required conveyors are shown in the processing area. Figure 5.4 is a preliminary layout of the RDF receiving, storage, and boiler plant to be located near the existing downtown district heating plant. Table 5.13 shows that the total capital requirement for the RDF processing plant and the boiler plant is about $97 million. Table 5.14 indicates a combined 1985 operating cost of about $17 million for a system with a design capacity of 8,650 TPW. Table 5.15 gives the labor schedule assumed for both facilities.

TABLE 5.10

River City Alternative 3 - Co-generation, Capital Costs (8,400 TPW)

Item	Capital Costs ($ x 1000)
1. Site Work	500
2. Buildings: raw refuse pit, incinerator/turbine building, and receiving building	9,310
3. Mechanical Equipment: pit cranes, scales, dust collection system, turbine room crane, and installation	2,670
4. Steam/Power Equipment: two, 800 TPD (188,000 pph steam) waterwall incinerators (60 psig/750 F.), pumps, fans, stacks, water treatment, other boiler auxillaries, one 35 MW extraction turbine, cooling tower, substation, and installation.	36,820
5. Rolling Stock: ash trucks, pick-up, and sweeper	290
6. Energy Distribution: steam line, and electric power line to utility tie-in-point	1,000
7. Subtotal (items 1-6)	50,590
8. Contingency (15% of item 7)	7,589
9. 1980 Construction Cost (items 7 & 8)	58,179
10. 1983 Connstruction Cost[a]	81,737
11. Eng., Legal, Admin. (10% of item 10)	8,174
12. Start-up Reserve[b]	2,335
13. Subtotal (items 10-12)	92,246
14. Financing Costs (30% of item 13)[c]	27,674
15. Total Capital (item 13 & 14)	119,920

[a] Escalation from 1980 $ to 1983 $ at 12% per year.
[b] Six months of 1985 O & M costs, see Table 5.11.
[c] Assumes revenue bond issue by the City.

TABLE 5.11

River City Alternative 3 - Co-generation, Annual Costs
(8,400 TPW)

Item	Annual Costs ($ x 1000)
1. Site and Bldg. Maintenance	50
2. Mechanical Equipment Maintenance	90
3. Steam/Power Equipment Maintenance	740
4. Labor[a]	1,200
5. Electric Power	800
6. Other Utilities	20
7. Ash Haul and Disposal[b]	540
8. Subtotal (items 1-7)	3,440
9. Contingency (10% of item 8)	344
10. 1980 O & M Cost (items 8 & 9)	3,784
11. 1985 O & M Cost[c]	6,094
12. Debt Service[d]	13,139
13. Total Annual Cost (items 11 & 12)	19,233

[a] 60 employees at an average annual salary, overhead, and fringe benefits cost of $20,000, see Table 5.12.
[b] Assume 46 miles round trip, 30% of raw refues.
[c] Escalation from 1980 $ to 1985 $ at 10% per year.
[d] 9%, 20-year financing of Total Capital from Table 5.10.

TABLE 5.12

River City Alternative 3 - Co-generation, Labor Schedule (8,400 TPW)

Job Category	No. per Shift	No. of Shifts	Total Required
Superintendent	1	1	1
Forman	1	4	4
Scale Operator	1	2	2
Traffic Director	1	2	2
Crane Operator	2	4	8
Boiler Operator	2	4	8
Turbine Operator	1	4	4
Plant Mech. Engineer	1	1	1
Plant Elec. Engineer	1	1	1
Ash Loader	2	4	8
Laborer	2	2	4
Mechanic	1	2	2
Electrician	1	2	2
Truck Driver	3	1	3
Relief Operators	2	4	8
Steno/Clerk	2	1	2
Total			60

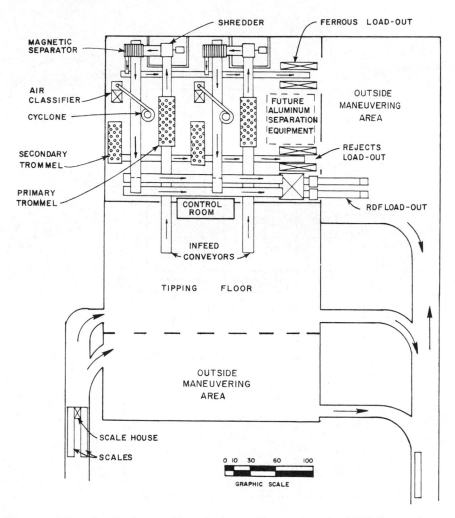

Figure 5.3. Preliminary Site Layout--Alternative 4, RDF Processing Plant

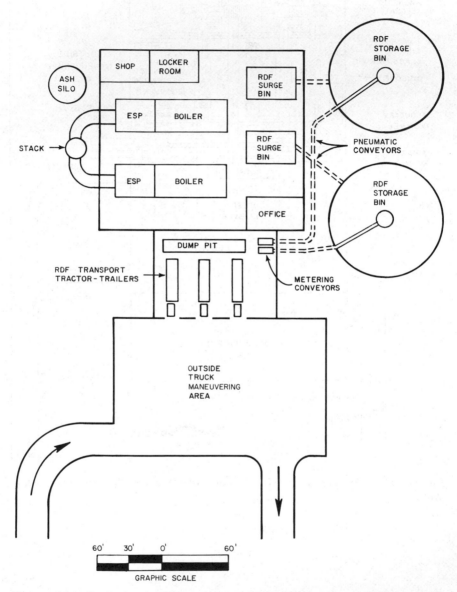

Figure 5.4.   Preliminary Site Layout--Alternative 4, Boiler Plant

TABLE 5.13

River City

Alternative 4 - Steam for District Heating, Capital Costs (8,650 TPW)

Item	Capital Costs ($ x 1000)
Transfer Station Modification Costs:	
1. Site Work	60
2. Buildings	160
3. Equipment:	7,060
Primary and Secondary Trommels, shredders, magnetic separators, air classification systems, dust collection systems, conveyors, other equipment and installation	
4. Rolling Stock:	780
Front-end loaders, residue trucks, RDF transfer tractor-trailers, pick-ups, and sweepers	
Boiler Plant Costs:	
5. Site Work	550
6. Buildings	5,210
7. Mechanical Equipment	2,180
8. Steam Equipment:	23,380
two, 200,000 pph steam boilers (650 psig, 750 F.), pumps, fans, water treatment, ash system, stacks, other auxillaries, installation	
9. Rolling Stock	290
10. Subtotal (items 1-9)	39,670
11. Contingency (15% of item 10)	5,951
12. 1980 Construction Cost (items 10 & 11)	45,621
13. 1983 Construction Cost[a]	64,094
14. Eng., Legal, Admin. (10% of item 13)	6,409
15. Start-up Reserve[b]	3,160
16. Purchase of Existing Transfer Station[c]	683
17. Subtotal (items 13-16)	74,346
18. Financing Costs (30% of item 17)[d]	22,304
19. Total Capital (items 17 and 18)	96,650

[a]   Escalation from 1980 $ to 1983 $ at 12% per year.
[b]   Six months of 1985 O & M costs see Table 5.14.
[c]   Equivalent to original construction cost.
[d]   Assumes revenue bond issue by the City.

TABLE 5.14

River City

Alternative 4 - Steam for District Heating, Annual Costs
(8,650 TPW)

Item	Annual Costs ($ x 1000)
RDF Processing Plant	
1. Site and Building Maintenance	50
2. Mechanical Equipment Maintenance	260
3. Labor[a]	1,000
4. Electric Power[b]	410
5. Other Utilities	20
6. Residue Haul and Disposal[c]	280
7. RDF Haul[d]	50
Boiler Plant	
8. Site and Building Maintenance	30
9. Mechanical Equipment Maintenance	70
10. Steam Equipment Maintenance	470
11. Labor[e]	440
12. Electric Power	310
13. Other Utilities	10
14. Ash Haul and Disposal[f]	160
15. Subtotal (items 1-14)	3,560
16. Contingency (10% of item 15)	356
17. 1980 O & M Cost (items 15 and 16)	3,916
18. 1985 O & M Cost[g]	6,307
19. Debt Service[h]	10,589
20. Total Annual Cost (items 18 & 19)	16,896

[a]   50 employees at an average of $20,000 per employee per year including overhead and fringe benefits, see Table 5.15.

[b]   40 KWH per ton processed.

[c]   26-mile round trip, 20% of raw waste, $3.00 per ton disposal.

[d]   Four-mile round trip, 75% of raw waste.

[e]   22 employees at an average of $20,000 per employee per year including overhead and fringe benefits, see Table 5.15.

[f]   26-mile round trip, 10% of raw waste, $3.00 per ton disposal.

[g]   Escalation from 1980 $ to 1985 $ at 10% per year average.

[h]   9%, 20-year financing of Total Capital from Table 5.13.

TABLE 5.15

River City

Alternative 4 – Steam for District Heating, Labor Schedule (8,650 TPW)

Job Category	No. per Shift	No. of Shifts	Total Required
RDF Processing Plant			
Superintendent	1	1	1
Forman	1	2	2
Scale Operator	1	2	2
Traffic Director	1	2	2
Front-end Loader Operator	2	2	4
Equipment Operator	2	2	4
Residue loader	2	2	4
RDF Loader	3	2	6
Laborer	2	2	4
Mechanic	1	2	2
Electrician	1	2	2
Residue Truck Drivers	3	1	3
RDF Truck Drivers	3	2	6
Relief Operators	3	2	6
Steno/Clerk	2	1	2
Total			50
Boiler Plant			
Foreman	1	4	4
Boiler Operator	2	4	8
Ash Loader	1	4	4
Truck Drivers	2	1	2
Relief Operators	1	4	4
Total			22

REFERENCES

1. Robert Snow Means Co., Inc., Building Cost Construction Data, 1980, 38th Annual Edition, R.S. Godfrey ed., R.S. Means, Kingston, Mass., 1979.

2. McGraw-Hill Cost Information Systems, 1980 Dodge Manual for Building Construction and Pricing, Annual Edition No. 15, P.E. Pereira ed., McGraw-Hill, New York, 1979.

3. Berger Cost File, Inc., 1980 Berger Building Cost File--Unit Prices, Central Edition, Van Nostrand Reinhold, Co., New York, 1980.

4. International Construction Analysts, The Richardson Rapid System, 1979-1980 Edition, Vols, 1-4, Richardson Engineering Services, Inc., Solana Beach, California, 1979.

5. U.S. Environmental Protection Agency, Draft Environmental Impact Statement on the Proposed Guidelines for the Landfill Disposal of Solid Waste, OSW, EPA, Washington, D.C., March, 1979.

6. Robert Snow Means Co., Inc., Mechanical and Electrical Cost Data, 1980 - 3rd Annual Edition, M.J. Mossman ed., R.S. Means, Kingston, Mass., 1980.

7. U.S. Environmental Protection Agency, Procedures Manual for Ground Water Monitoring at Solid Waste Disposal Facilities, EPA/530/SW-611, Washington, D.C., August, 1977.

8. Kidder, Peabody & Assoc., Official Statement, Pinellas County Resource Recovery Project, New York, July 1, 1978.

9. Builders' Construction Cost Indexes, Engineering News Record, Vol. 204, No. 25, June 19, 1980.

10. Salomon Brothers, Inc., City of Phoenix Resource Recovery Project--Financial Feasibility, December, 1979.

11. U.S. Environmental Protection Agency, Resource Recovery Plant Implementation, Guides for Municipal Officials--Financing, Robert E. Randol, ed., SW-157.4, Cincinnati, Ohio, 1975.

12. E.L. Grand and W.G. Ireson, Principles of Engineering Economy, Ronald Press, New York, 1970.

13. Bruce E. Fritch and Albert F. Reisman, Editors, Equipment Leasing--Leveraged Leasing, Practising Law Institute, New York, 1977.

14. Kathryn Grover, "Equipment Leasing: The Newest Boom in Government Purchasing," in American City and County, May 1979, p. 83.

CHAPTER 6

SYSTEMS ANALYSIS AND LIFE CYCLE COSTS

I. INTRODUCTION

The previous chapters of this book have dealt with the collection of
basic data, the identification of markets, the selection of appropriate
system alternatives, and the calculation of costs for facilities.   The
purpose of this chapter is to present a methodology for combining all
of this information in an economic analysis structure for use as a tool
in decision-making.   The methodology presented in this chapter can be
used to decide for each alternative 1) the most economical locations
for facilities, 2) the most economical facility capacities and resulting
service areas, and 3) the proper number of facilities in multi-facility
systems.   The methodology can also be used to illustrate the magnitude
of the net cost differences between alternative systems both for the
initial operating year and over the system's operating life.   The
methodology also allows the investigator to conduct sensitivity analyses;
that is, test the effects of changes in certain assumptions (inflation
rates, interest rates, solid waste quantity, energy and materials unit
revenues, etc.) on the net system cost of each alternative.

Section II of this chapter details a mathematical modeling
technique for system alternatives which is used to optimize the facility
configuration of each alternative and compare initial operating year
system costs.   Section III presents a life cycle cost analysis technique
which is used for comparing system costs over the operating life of the

system as well as providing a structure for sensitivity analysis. The final section of this chapter continues the River City example for numerical illustration.

II. SYSTEM MODELING

The objective of system modeling is to provide a method for simulating the costs of alternative solid waste management systems to identify the lowest cost alternative.

A model is used which describes the different solid waste management system alternatives mathematically so that the system costs can be assigned to various elements of the system. The mathematical model is then used to select the configuration of facilities and waste assignments to these facilities which minimize the study area's net costs for transportation, processing/combustion, and disposal. The net system cost (NSC) is composed of all of the costs of operating the alternative in question starting with transportation after collection through final disposal minus all revenues which may be accumulated.

Figure 6.1 graphically illustrates the potential options available for disposal of waste which originates in a waste generation district (WGD). The WGD can transport waste directly to a transfer station, resource recovery facility or landfill. From the transfer station, waste can be transported to a resource recovery facility or landfill. The resource recovery facility can produce a fuel (or another energy product) and/or a recovered material which is sent to a buyer. A residue may also be produced which is transported to a landfill. Each transportation link and every facility has an associated cost of operation or utilization.

Figure 6.1 illustrates the combinations available to a single WGD. Given a study area with hundreds of WGDs, many potential transfer station sites and sizes, and several potential resource recovery facility locations, the possible combinations of WGD waste assignments and the possible combinations of facility sizes and locations are legion. Even

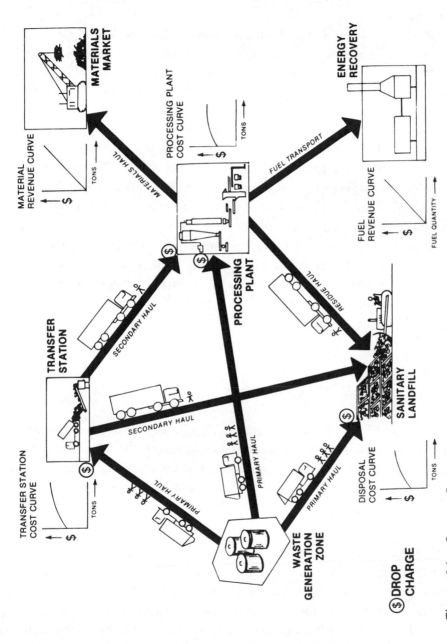

Figure 6.1.   System Components

for a modestly complicated problem, the number of combinations can be overwhelming.

For this reason, system modeling problems have been reduced to a set of equations and solved using computers. The detailed mathematical formulation of this type of model has been described in the literature (1-6). The investigator should consult these references for details. Linear programming techniques are used to solve the equations (7,8). Several commercially available computer software packages have the capability to solve these kinds of linear programming problems. It is the purpose of this section to present the concept of this type of mathematical modeling while avoiding detailed descriptions of systems analyses procedures. The generalized form of equations presented could be solved by hand.

The NSC for each alternative has the following components: transport costs, facility costs, and revenues. The following paragraphs illustrate mathematical representations of each of these components, and the resulting system model.

A. Transport Costs

The two basic types of transport costs (primary haul and secondary haul) have been discussed previously in Chapter 2. Mathematical representations for each type are discussed in the following paragraphs.

1. Primary Haul

Primary haul costs are incurred in transporting raw waste from the end of the collection route to a landfill, transfer station, or a resource recovery facility (costs for transport during collection are not included). In order to calculate these costs, the waste quantities per unit time (tons/yr, tons/wk, tons/day) by WGD must be utilized along with transport time and distance between each WGD and the unit haul costs for collection vehicles representative of each WGD. The methods for developing these data were discussed in Chapter 2. Note that the formulation presented here utilizes a "composite vehicle" unit haul cost as presented in Chapter 2. If individual haul costs for different waste types is to be used, the model becomes more complicated.

For any particular landfill, resource recovery facility, or transfer station, the primary haul cost is the sum of the primary haul costs from each WGD contributing waste to the facility. Using the basic data previously gathered, this relationship can be stated mathematically. For any facility j,

$$P_j = \sum_{i=1}^{I} \left[\left(\frac{PMC_i \; D_{ij} \; W_{ij}}{C_i} \right) + \left(\frac{PHC_i \; T_{ij} \; W_{ij}}{C_i} \right) \right] \qquad (1)$$

where:

P_j =   Primary haul cost to facility j in dollars per unit time.

i =   Individual WGD i.

j =   Individual facility j.

I =   Total number of WGDs in the analysis.

PMC_i =   Unit haul cost in \$/mile for WGD i.

D_{ij} =   Round-trip distance in miles between WGD i and facility j.

W_{ij} =   Amount of waste transported from WGD i to facility j in tons per unit time.

C_i =   Capacity of the collection vehicle for WGD i in tons.

PHC_i =   Unit haul cost in \$/hour for WGD i.

T_{ij} = Round-trip travel time in hours between WGD i and facility j plus unloading time at facility j.

Note that the equation can be simplified considerably if a composite per-mile unit haul cost is utilized instead of separate per-mile and per-hour unit haul costs.

Note that because private collectors are normally paid for collection, transport, and disposal combined, and because standard accounting practices for municipal collection services do not specifically display primary haul costs, these costs are not normally budgeted as a separate item for solid waste management. They are, nevertheless, actual operating costs (whether incurred by a private collector or a municipal collector) which vary with facility location. As such, these costs should be a factor in a comparative economic analysis of systems which have different facility locations.

Equation (2) is also subject to certain constraints. First, the sum of all waste transported to any facility must not exceed that facility's input capacity. Mathematically,

$$\sum_{i=1}^{I} W_{ij} \leq CAP_j \qquad (2)$$

where CAP_j = the input capacity of facility j.

For instance, if a resource recovery facility or a transfer station has the capacity to handle 2,500 tons per week, the system model must not assign more than this capacity to the facility.

In addition, all waste generated in any WGD must be transported to one or more facilities. Mathematically,

$$W_i = \sum_{j=1}^{J} W_{ij} \qquad (3)$$

where W_i = all waste generated in WGD i per unit time, and J = total number of facilities in the analysis.

2. Secondary Haul

Secondary haul costs are incurred in hauling raw waste from a transfer station to a landfill or resource recovery facility, or residue and/or ash from a resource recovery facility to a landfill. Costs for transporting recovered materials or energy products to the ultimate buyer may also be considered secondary haul costs, but many times these costs are either included as facility operating costs, or are subtracted from revenues to produce a net revenue. Secondary haul costs must be included in the analysis because these costs may influence the location of transfer stations (and in some cases resource recovery facilities).

Secondary haul costs are calculated by utilizing the unit haul costs for transfer vehicles and residue/ash vehicles along with the time and distance information between the facilities. Methods for developing these data were discussed in Chapter 2. For a particular system, the secondary haul cost is the sum of the costs of transporting raw waste or other materials between each facility in question. Mathematically, this relationship can be represented as,

$$S_j = \sum_{m=1}^{M} \left[\left(\frac{SMC_{jm} \; D_{jk} \; W_{mjk}}{C_{jm}} \right) + \left(\frac{SHC_{jm} \; T_{jk} \; W_{mjk}}{C_{jm}} \right) \right] \qquad (4)$$

where:

S_j =   Total secondary haul cost for facility j in dollars per unit time.

m =   Type of material (raw waste, residue/ash or recovered materials).

j =   Origin facility.

SMC_{jm} =   Unit haul cost in $/mile for facility j and material type m.

k   = Destination facility.

D_{jk} = Round-trip distance in miles between facility j and facility k.

W_{mjk} = Amount of material type m transported between facility j and facility k per unit time.

C_{jm} = Capacity of vehicle hauling material type m from facility j in tons.

SHC_{jm} = Unit haul cost in \$/hour for facility j and material type m.

T_{jk} = Round-trip travel time in hours from facility j to facility k plus unloading time at facility k.

Note that when facility j is a transfer station, $SHC_{jm} = 0$ because driver labor is included in the operating cost of the transfer station (see Chapter 2).   Also when facility j is a transfer station, all waste entering the facility must be transported out.   Mathematically,

$$\sum_{i=1}^{I} W_{ij} = \sum_{k=1}^{K} W_{jk} \tag{5}$$

Note also that if facility j is a transfer station, only one material type is transported (solid waste).   However, if facility j is a resource recovery facility, more than one material may be involved (residue, ash, or recovered materials).

B.   Facility Costs

Facility cost estimating techniques were discussed in Chapter 5.   Refer to this previous discussion for guidance in estimating the costs which are the input data to this analysis. For the purposes of this mathematical model, there are two components to the facility cost: operation and maintenance (O&M), and debt service (or amortized capital). Both are expressed in dollars per unit time (dollars/day, dollars/week, dollars/year, etc.)

O&M costs include maintenance, labor, utilities, and other costs. Some O&M costs are fixed with respect to the amount of waste handled by the facility for a given facility capacity. For example, site and building maintenance, a certain portion of labor, and insurance can be termed "fixed" costs. The other O&M costs vary directly with the amount of waste handled by the facility (e.g. electric power, equipment maintenance). The total O&M cost for a given facility can be calculated by adding the fixed and variable costs for a given waste quantity throughout. Mathematically,

$$OM_j = FC_j + VC_j \sum_{i=1}^{I} W_{ij} + \sum_{k=1}^{K} W_{kj} \qquad (6)$$

where:

OM_j = Operation and maintenance cost for facility j in dollars per unit time.

FC_j = Fixed costs for facility j in dollars per unit time.

VC_j = Variable costs for facility j in dollars per ton.

W_{kj} = Amount of waste transported from facility k to facility j in tons per unit time.

K = Total number of intermediate facilities (transfer stations).

Debt Service for a particular facility is a function of financing method, interest rate, and total capital cost. See Chapter 5 for an explanation of calculation methodology. This facility cost component does not vary with the amount of waste handled by the facility; that is, it is a fixed cost. For the purpose of this mathematical model, let A_j represent the debt service cost in dollars per unit time for any facility j.

Note that the disposal costs for residue and/or ash should be included in the variable O&M cost by utilizing the factors given in Table 4.4, Chapter 4. Disposal costs for non-processable wastes and excess processable waste (in the case of an energy market demand capacity limitation) are included in the analysis by including the O&M and debt service costs for a properly-sized sanitary landfill as one of the required facilities.

C. Revenues

The revenue component of net system cost acts to offset the other two components when a resource recovery facility is part of the system under consideration. Revenues are mostly derived from the sale of recovered energy and materials. The methodology for identifying potential buyers for these recovered resources and the market value of each was presented in Chapter 3. Revenues may also come from the investment of certain bond reserve funds which must be retained over the life of the facility as a condition of certain types of financing. For a particular resource recovery facility j, the total revenues are the sum of the amount of the recovered resource multiplied by the appropriate market value for all recovered resources and buyers, plus the amount of fixed revenue from investment earnings. Mathematically,

$$R_j = FR_j + \sum_{z=1}^{Z} \sum_{x=1}^{X} (RR_{xjz} \, V_{xjz}) \qquad (7)$$

where:

R_j = Total revenue income to facility j in dollars per unit time.

FR_j =   Fixed revenue from investment earnings at facility j.

RR_{xjz} =   Amount of recovered resource of type x sold by facility j to market z in units appropriate to the type of recovered resource (i.e., tons/day, pounds per hour, KWH per month, etc.).

V_{xjz} =   Unit market value for recovered resource x sold by facility j to market z in appropriate units (i.e., dollars/ton, dollars/1000 lb., dollars/KWH).

Note that in this analysis, "drop charges" or "tipping fees" which may be collected at existing or future landfills, transfer stations, or resource recovery facilities are not considered sources of revenue for the purpose of this mathematical model. Funds from such fees collected by a municipality at solid waste facilities may be considered "revenue" from the standpoint of the municipal budget, however, such fees are ultimately the burden of the taxpayer or ratepayer in the study area. These fees are not, therefore, "revenues" from the standpoint of net system costs to the community.

D.   Net System Cost

The NSC for any particular configuration of facilities and waste assignments, is the sum of the transport and facility costs minus the revenues for all facilities in the particular configuration under consideration.   Mathematically,

$$NSC = \sum_{j=1}^{J} (P_j + S_j + OM_j + A_j - R_j) \qquad (8)$$

When utilizing equation (9), the investigator should assure that all components of the NSC for a particular facility j are calculated on the basis of the same "base year." The base year normally utilized should be the first full year of operation of the facility in question. For example, if the facility costs (OM_j and A_j) are calculated on the basis of a 1985 start-up year, transport costs (P_j and S_j) and revenues (R_j) should be adjusted utilizing appropriate inflation factors to the same 1985 base year.

The investigator should also assure that the units of each component of NSC are the same when using equation (9). Units of dollars per day, dollars per week, and dollars per year can be used. Using units of dollars/day is not recommended, however, because different facilities in the same system configuration may operate for a different number of days per week causing problems in calculation consistency.

The investigator should not expect this mathematical model to cause the "optimum" solid waste management system to miraculously appear out of a jumble of data. It is a tool to provide information to the investigator which is not apparent without the structure of the model. It provides a method for quickly investigating the cost implications of different system configurations so that informed judgement can be made. The accuracy or applicability of the result of system modeling (the net system cost) is sensitive to the accuracy of the input data (waste quantities, population projections by WGD, employment projections by WGD, transportation time and distance between WGDs, facility costs, and potential facility sites). The investigator's confidence in the accuracy of these input data should be reflected in his confidence in the accuracy of the results of the system modeling analysis.

III. LIFE CYCLE COST ANALYSIS

The result of the system modeling analysis is a net system cost (dollars per week or dollars per year) in a "base year" (usually the first full year of operation) for each alternative system. The investigator can select the system with the lowest base year NSC for implementation, but to provide an additional level of analysis, the calculations must be carried one step further. Changing costs of operation due to inflation along with changing market values for recovered energy and materials will ensure that the net system cost calculated for each system in the base year will not remain static, but will change over time. From a municipality's point of view, it is difficult to make a sound decision on the feasibility of a multi-million dollar resource recovery project based on comparing the costs in a single initial operating year with the costs of landfilling or other alternatives. The NSC of alternative systems may change relative to one another over the study period. This means that the lowest cost alternative initially may not retain this ranking over the life of the facility.

Life cycle costing is an economic analysis tool which attempts to account for both the initial purchase price of a facility and the fixed and variable operating costs over the anticipated operating life. Life cycle costing methods were developed in recognition of the premise that decisions should not be based solely on initial cost considerations. The principles of life cycle costing have traditionally been applied in the construction industry as tools for owners to decide on the most cost-effective building materials, lighting systems, mechanical systems, and insulation systems, for example. These principles apply equally well for making decisions about the economic feasibility of solid waste resource recovery projects. The purpose of this section is to present the principles of life cycle costing as applied to resource recovery projects and to illustrate how the investigator can use the resulting information in the decision-making process.

A. Data Requirements

Simply stated, the life cycle costing (LCC) methodology starts with the elements of net system cost (transport costs, facility costs, and revenues), inflates them over the study period, adjusts these projected costs for the time value of money, then sums the resulting adjusted

annual costs. The data required for such calculations includes: the
base year NSC cost elements, inflation rates, and discount rates (time
value of money).   Also, because LCC provides a structure for
sensitivity analysis, the parameters to be used in the sensitivity analysis
should be identified in advance.   Note that because of the inherent
inaccuracies in assumptions of inflation rates, discount rates, and the
NSC cost elements, the results of the LCC analysis may not reflect
actual life cycle costs.   Comparisons between alternatives (if
assumptions are consistent) can, however be made with a certain
measure of confidence.

1. Base Year NSC Cost Elements

The previous section of this chapter explains the methodology for
calculating transport costs, facility costs, and revenues.   The
investigator should refer to Section II of this chapter for details.   A
few additional considerations are, however, necessary for LCC analysis.
The elements of facility operation and maintenance (O&M) costs were
presented as either "fixed" or "variable" with respect to waste quantity.
However, most of the "fixed" cost items are subject to annual cost
escalation due to inflation (building & site maintenance, insurance,
certain labor costs, etc.).   Other "fixed" costs such as land lease may
not be subject to annual inflation.   The investigator must decide which
costs are subject to inflation and which are not.   Generally, variable
costs are subject to cost escalation as a result of both inflation and
the growing amount of solid waste processed or handled by the
facilities in question.

In addition, certain "fixed" revenues (investment income from
certain bond reserve funds) may be subject to arbitrage regulations
limiting the investment interest rate, thereby resulting in a revenue
source which is not affected by inflation.   The investigator must make
this determination in advance of performing the LCC analysis.   Most
revenues, especially energy products, will be subject to escalation due
to inflation, and the rising amount of solid waste generated, unless the
facility is capacity-limited by energy market demand.

2. Inflation Rates

The inflation rate indicates a change in the general level of prices or monetary purchasing power through time. Inflation factors must be chosen for the various components of NSC in order to perform an LCC analysis. There are two ways of approaching the selection of inflation rates (8). First, an average inflation rate can be assumed for all components of NSC, or for individual components and applied to the appropriate cost and revenue items. For example, the investigator may wish to assume a 7 percent annual rate for labor, 10 percent for energy (costs and revenues), and 5 percent for all other items. Predicting the long-term average inflation rate for different cost and revenue items is certainly a speculative exercise. Estimates can, however, be made by utilizing regression analysis on various cost indices (10), and complex models have also been utilized to predict the changing cost of energy relative to other products by considering U.S. energy reserves, world energy reserves, international politics, etc. (11). The investigator can, therefore, make "reasonable" inflation assumptions. These assumptions can also be varied in a sensitivity analysis to identify the overall effect on the feasibility of a resource recovery project.

The second approach is to escalate only the energy costs such as electric power and auxiliary fuel, and revenues. A differential escalation rate can be defined for energy which is the difference between the expected average inflation rate and the expected energy inflation rate. This is the method which has been selected in federal government LCC analyses (12-14). Although it may appear simpler to utilize this method, the determination of a differential energy inflation rate involves the same degree of uncertainty as the first method.

3. Discount Rates

The discount rate and inflation rate are two distinct factors. The discount rate is the interest rate which reflects the time value of money and is used to convert expenditures and revenues occurring at different times to a common time frame. The federal government utilizes a "lost opportunity" concept in determining the discount rate to be used in federal LCC analyses. The federal Office of Management and Budget has set a rate of ten percent per year <u>above</u> the annual inflation rate as the proper discount rate (15). It is argued that the federal government, which ultimately depends on taxes to function, diverts funds used in federal projects from private investment, therefore creating a "lost opportunity" for investment by private industry. It is further argued that ten percent per year represents the average return on investment (after adjusting for inflation) for private industry.

This concept can also be applied on a municipal level. References 16 through 18 should be consulted for further information on this approach to the determination of discount rates.

There is, however, another approach to determining the discount rate for a resource recovery analysis under certain conditions. A popular method of financing resource recovery projects is via tax-exempt revenue bonds (see Chapter 5). Unlike general obligation bonds, the municipality does not obligate its tax base to the repayment of the debt. The revenues from energy and recovered materials sales as well as disposal charges or drop fees are, instead, pledged to repay the bond holders. Therefore, unlike traditional "public works" type projects (e.g., streets, sewers) this type of financing does not limit the municipality's capacity to borrow funds for other projects nor does it directly draw on the tax base. It can be argued that under these financing conditions, there is no "lost opportunity" by the diversion of funds from the private sector either through solid waste tipping fees or energy sales. This is because if the project is economically feasible, tipping fees to the taxpayer will be equal to, or less than existing (or projected) landfilling fees and the energy buyer will be paying for recovered energy at rates equal to or less than the costs of energy from "traditional" sources. The conclusion is that for a municipally-owned resource recovery project financed with tax-exempt revenue bonds, the discount rate should be equal to the bond interest rate. In either approach it is important to select a discount rate which appropriately corresponds to the type of inflation rate selected--total or differential.

4. Sensitivity Analysis Parameters

One of the benefits of LCC analysis is that the life cycle cost implications of changes in assumptions can be quickly tested. For example, changes in discount rates, inflation rates, or energy market values can be made to test the extent of the resultant changes in life cycle costs for alternative systems. The major candidate parameters for the sensitivity analysis should be identified before the LCC analysis is performed so that the basic data affected by changing these items are readily identified.

B. Calculation Methodology

There are two basic ways to express life cycle cost: 1) the present worth (PW) method, and 2) the uniform annual cost (UAC) method. The PW method reduces the net costs of a particular alternative over its operating life into a single, lump sum. The UAC method calculates a theoretical equivalent annual dollar amount which is uniform over the life of the facility. Either method produces a number which can be used to compare alternative systems.

For the PW method, the calculations begin with determining the present worth of the accumulated debt service. This is accomplished by multiplying the annual debt service by the uniform series present worth factor for the discount rate d and time span N. Mathematically, for facility j,

$$PA_j = A_j \ \frac{(1+d)^N - 1}{d(1+d)^N} \tag{9}$$

where:

PA_j = Accumulated present worth of the series of debt service payments over the life span N of facility j in dollars.

d   = Annual discount rate.

N   =   Facility life span in years.

Note that tables are available for the expression $(1+d)^N - 1 \ / d \ (1+d)^N$ for various rates and time periods in numerous reference materials (19). Note also that if the discount rate is equivalent to the borrowing rate, this calculation need not be performed because PA_j will be equivalent to the total capital cost in this circumstance.

The next step is to calculate the accumulated present worth of the fixed O&M costs. Those fixed O&M costs not subject to inflation are multiplied by the uniform series present worth factor, as discussed previously for debt service. Fixed O&M costs which will escalate due to inflation must first be adjusted for the inflation rate, then multiplied by a present worth factor which accounts for the discount rate. This can be accomplished by multiplying the single payment compound amount factor by the single payment present worth factor for each individual year in the life span. Tables for these factors are available in numerous references (18).

The summation of each year's factor multiplied by the fixed annual costs which are subject to inflation plus the present worth of the fixed O&M costs not subject to inflation gives the required result. Mathematically, for facility j,

$$
FOM_j = NFC_j \frac{(1+d)^N - 1}{d(1+d)^N} + IFC_j \sum_{n=1}^{N} \frac{(1+c)^n}{(1+d)^n} \qquad (10)
$$

where:

FOM_j = Accumulated present worth of the fixed O&M costs for facility j in dollars.

NFC_j = Fixed annual costs for facility j not subject to inflation in dollars per year.

IFC_j = Fixed annual costs for facility j which escalate with inflation in dollars per year.

c   = Annual inflation rate.

n   = Year of analysis.

Note that tables exist (20) for the expression:

$$\sum_{n=1}^{N} (1+c)^n/(1+d)^n$$

Note also that the sum of the non-inflated and inflated fixed costs equals the total fixed cost of the facility, or $NFC_j + IFC_j = FC_j$. Note also that equation (11) illustrates the use of a composite average inflation rate (c). If different inflation rates are to be used for different fixed cost components, the equation becomes more complicated.

The accumulated present worth of the variable facility O&M costs are calculated next. This calculation is accomplished by multiplying the variable cost (in dollars/ton) by the waste quantity handled by the facility in question in year n. Adjustments for inflation and discount rates are made in each year as discussed previously, then the results are summed over the facility life span. Mathematically, for facility j,

$$VOM_j = VC_j \sum_{n=1}^{N} W_{jn} \frac{(1+c)^n}{(1+d)^n} \qquad (11)$$

where:

VOM$_j$ = Accumulated present worth of the variable O&M costs of facility j in dollars.

$$W_{jn} = \sum_{i=1}^{I} W_{ij} + \sum_{k=1}^{K} W_{kj} \quad \text{in year n (in tons).}$$

Note again, that equation 12 illustrates the required calculation when a composite inflation rate is used. If separate inflation rates for different variable cost components are used, the equation becomes more complicated because separate calculations are required.

The accumulated present worth of the transport costs are calculated in much the same manner,

$$PT_j = P_j + S_j \sum_{n=1}^{N} \frac{(1+c)^n}{(1+d)^n} \qquad (12)$$

where:

PT_j = Accumulated present worth of transport costs for facility j in dollars.

Note that equation 13 takes into account changes in transport costs due to inflation and adjusts the costs with a discount rate, but does not account for cost changes due to future changes in the transportation network (e.g., new roads, widening of certain roads to reduce travel time. Because it is possible that the transportation network will change in some study areas significantly over a 20- to 25-year study period, a more sophisticated LCC analysis utilizing a time-staged transport cost can be made. Projected changes in the transport network can be used to generate updated primary haul costs (P_j) and secondary haul costs (S_j) at intervals during the study period. Then equation 13 can be used to calculate an updated PT_j at appropriate intervals. However, this refinement of the model is not recommended for the feasibility study level of analysis unless major changes in the transport network are anticipated.

The calculation of the accumulated present worth of the fixed and variable revenues closely parallel the calculations for fixed and variable O&M costs. The equations are:

$$PFR_j = NFR_j \frac{(1+d)^N - 1)}{d(1+d)^N} + IFR_j \sum_{n=1}^{N} \frac{(1+c)^n}{(1+d)^n} \qquad (13)$$

and

$$PVR_j = \sum_{n=1}^{N} VR_{jn} \frac{(1+c)^n}{(1+d)^n} \tag{14}$$

where:

PFR_j = Accumulated present worth of the fixed revenues to facility j in dollars.

NFR_j = Fixed annual revenues not subject to inflation for facility j in dollars per year.

IFR_j = Fixed annual revenues which escalate with inflation in dollars per year.

PVR_j = Accumulated present worth of the variable revenues to facility j in dollars.

VR_{jn} = Total variable revenue from all sources for facility j in year n, or

$$\sum_{z=1}^{Z} \sum_{x=1}^{X} (RR_{xjr} V_{xjr})$$

Note that equation 14 is similar to equation 11, and equation 15 is similar to equation 12. Again, equations 14 and 15 illustrate the use of a composite inflation rate for all revenue sources. If separate inflation rates for energy revenues and recovered materials revenues, for example, are to be used, separate calculations for each are required.

The total LCC present worth is the sum of equations 10,11,12 and 13 minus the sum of equations 14 and 15 for all facilities in the alternative system under consideration,

$$PW = \sum_{j=1}^{J} FOM_j + VOM_j + PT_j - PFR_j - PVR_j \tag{15}$$

where PW = total accumulated present worth of the system over the study period. PW can be calculated for all of the alternative systems under examination to see how they rank with respect to life-cycle costs, compared with how they rank on a base-year NSC basis.

Because in many resource recovery projects the annual waste quantity varies in each operating year (and hence costs and revenues vary), the uniform annual cost (UAC) calculation for LCC can only be accomplished after determining PW with equation 16. To determine UAC, the PW for a given alternative system is simply multiplied by the capital recovery factor for the discount rate d and the life span N. Mathematically,

$$UAC = PW \; \frac{d(1+d)^N}{(1+d)^N - 1} \tag{16}$$

where UAC = uniform annual life cycle cost (dollars/year). Note that tables are available in numerous references for capital recovery factors (19). The resulting number in dollars per year can be used to compare alternative systems in much the same manner as the PW method. The advantage of calculating a UAC number for each alternative is that it gives the decision-makers a "feel" for the average annual net cost of a given alternative system over its operating life span in contrast to the lump-sum PW method.

C. Sensitivity Analysis

LCC analysis for resource recovery alternatives is a natural application for computer programming because of the number of calculations required. Generalized life cycle cost software is commercially available and can easily be applied to these types of problems. The investigator should be certain that the software selected has enough flexibility to allow for certain unusual aspects of resource recovery LCC analysis (such as variable waste quantities).

Once an LCC computer model of a particular system has been programmed, a sensitivity analysis for a wide range of variables can be quickly and easily obtained by changing one or more parameters in the analysis and re-running the computer simulation. The major candidate parameters for sensitivity analysis are: cost inflation rates; energy production rates, to examine the effects of different equipment efficiencies; energy and material market value; financing interest rate; and waste quantity. Because a significant degree of uncertainty exists in the determination of these pieces of data, it is useful to know the life cycle cost differences associated with a range of assumptions.

One way to utilize sensitivity analysis is to postulate a "worst case" and "best case" for each alternative to observe how the LCC rankings may change. The sensitivity analysis can also be a useful tool in negotiating with an energy market for the sale of recovered energy. With a computer model and a portable terminal, the effects of changes in the energy market value (dollars/kWh, dollars/1000 lb., etc.) on the life cycle costs to the community can immediately be observed.

IV. RIVER CITY ANALYSIS

The following economic analysis example for River City is the culmination of all of the example data gathered in previous chapters of this book. First, the base-year NSC costs are presented for each of the four alternatives, then the LCC analysis is presented.

A. Base-Year NSC

Table 6.1 is a summary of the base-year net system costs calculated for each alternative using equations similar to the ones given in Section II of this Chapter. Note that all of the costs displayed on Table 6.1 are in terms of 1985 dollars. The 1985 facility costs calculated in

TABLE 6.1
River City
Base-Year (1985) Net System Cost Components ($/yr. x 1000)

Alternative	Primary Haul	Secondary Haul	Facility Costs	Revenues	Net System Cost
1. New Landfill	3,975	1,750	4,723	0	10,448
2. Steam to RCI	3,459	472	16,605	8,824	11,712
3. Co-Generation	3,459	472	20,151	9,322	14,760
4. Steam for District Heating	3,420	408	17,598	5,412	16,014

Chapter 5 were utilized along with transport costs and revenues escalated to 1985 from 1980 cost calculations. Transport costs were escalated at 7 percent per year, recovered ferrous metals at 3 percent per year, steam revenues at 12 percent per year, and electric power revenues at 9 percent per year. Note also that the secondary haul costs shown are for raw waste only. Residue/ash haul costs and RDF haul costs (Alternative 4) are included in the annual facility costs.

It should also be recognized that the net system costs displayed on Table 6.1 do not include the costs for disposing non-processable wastes or excess processable wastes. Since calculations showed that existing landfills (mainly the North landfill) had sufficient remaining capacity for these wastes over the study period, no new landfill capacity was needed to support the resource recovery alternatives. Therefore, the costs of operating the existing landfills would be incurred in each Alternative. For this reason, these costs were removed from the facility costs in each alternative to simplify the analysis.

1. Alternative 1--New Landfill

The net system cost of all possible combinations of landfill locations (LF1 and LF 2), the five transfer station locations (see Figure 2.2, Chapter 2), and the three potential transfer station sizes were calculated. These calculations were preformed with a computer program. The configuration with the lowest NSC for this alternative is shown in Figure 6.2. It includes utilizing landfill site LF2 ten miles south of the study area. All waste is hauled via two 3750 TPW transfer stations at locations T2 and T3 (no direct haul to the landfill).

The facility costs in Table 6.1 for Alternative 1 include the two transfer stations ($745,000 per year per station) plus the operation of the new landfill ($3,233,000 per year). The facility costs include debt service as well as fixed and variable O&M. No revenues are generated in this alternative.

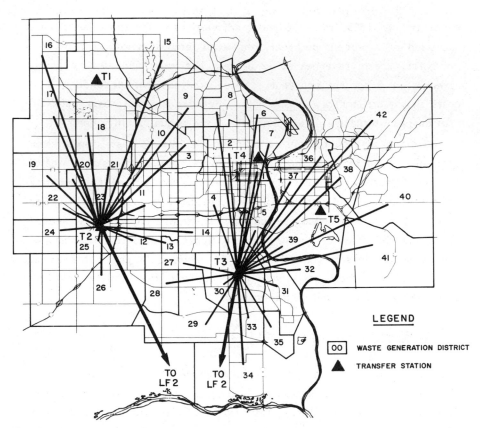

Figure 6.2.   River City Alternative 1 Configuration

2.   Alternative 2--Steam to RCI

A computer was again programmed to make NSC calculations for all possible combinations of four transfer station locations (T5 was eliminated because of its close proximity to the resource recovery plant site) and three transfer station sizes for the steam plant located at site RR2 (East Side Industrial Park).   The configuration with the lowest NSC is shown in Figure 6.3.   It includes a 2500 TPW transfer station at location T2 and a 1250 TPW transfer station at location T3.   About half of the study area's waste generation goes through the transfer stations to the resource recovery facility and half is transported directly.

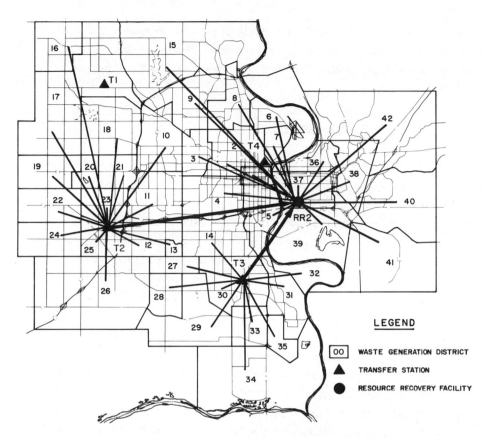

Figure 6.3.   River City Alternatives 2 and 3 Configuration

The facility costs for Alternative 2 include the annual costs for the 1250 TPW transfer station ($351,000 per year), the 2500 TPW transfer station ($567,000 per year) and the steam plant ($15,687,000 per year). Steam revenues were based on a market value of $2.50 per thousand pounds (Mlb) in 1980 escalated at 12 percent per year to 1985 to yield $4.40 per Mlb.   Based on the sale of about 2×10^9 pounds of steam per year (300,000 pph daily average, 24-hour per day, 6 days per week operation, 2 weeks per year RCI plant shutdown for scheduled maintenance), the 1985 annual steam revenue is about $8,660,000. Ferrous metals recovered from incinerator ash total 28,300 tons in 1985.   With a 1980 market value of $5.00 per ton escalated at 3 percent per year to $5.80 per ton in 1985, the metals revenue totals per year each) and the RDF processing plant and downtown boiler plant costs ($16,896,000). Because of lower steam demand during the summer about $164,000 per year. The revenues shown on Table 6.1 include both revenue sources.

3.   Alternative 3--Co-generation

Since the resource recovery facility site is the same in this alternative as for alternative 2 (RR2), the same configuration was used to calculate NSC for this alternative system.   Table 6.1 shows that as a result, the primary and secondary haul costs are the same as for Alternative 2.   The facility costs are higher because the co-generation plant has a higher annual cost than the steam plant ($19,233,000).   The total revenues are also higher because of the additional electric power revenues.   Electric power is generated mainly during the periods in which the RCI plant is not operating (one day per week).   Although there will be a space heating load during this day in the winter season, this load is small compared to the process steam load which will not be required.   This means that essentially the entire output of the steam plant can be used to generate electric power one day per week which results in sales (in 1985) of about 30 million kWh at an average load of 3,430 kW.   Using the 1980 market value of 8.4 mills per kWh and $1.75 per kW per month escalated at 9 percent per year to 12.9 mills per kWh and $2.70 per kW per month, the annual electric power revenue totals about $498,000.   Added to the steam revenue and metals revenue previously calculated for Alternative 2, the total revenues for Alternative 3 are $9,322,000 per year.

4.   Alternative 4--Steam for District Heating

A computer program was once again used to calculate the NSC for all possible combinations of four transfer station sites (T4 was eliminated because the processing plant is at this site), three transfer station sizes, and the RDF processing plant located at site RR1.   The configuration with the lowest NSC includes two 1250 TPW transfer stations located at sites T2 and T3.   About 60 percent of all waste generated is transported directly to the RDF processing plant (see Figure 6.4).

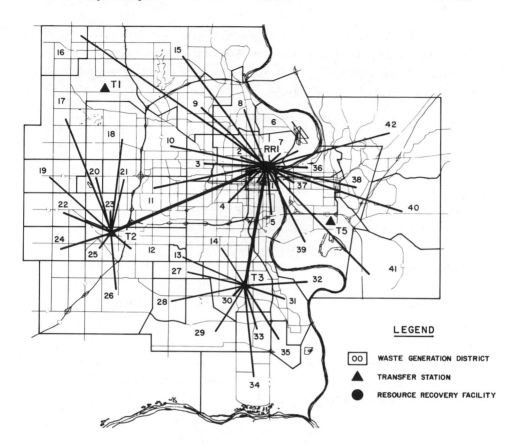

Figure 6.4. River City Alternative 4 Configuration

months, only about 60 percent of the steam which could be generated
from the RDF produced from the waste available in 1985 can be sold.
The remainder of the processable waste must be landfilled.   Utilizing
the same 1985 market value for steam as used for Alternatives 2 and
3, 1985 steam revenues would be about $4,920,000 per year.   Metals
revenue was calculated using a 1980 market value of $15.00 per ton
escalated at 3 percent per year to $17.40 per ton to yield about
$492,000 per year.

5. Summary of Base-Year NSC

Table 6.1 shows that on a 1985 net system cost basis, a new landfill
has the lowest cost. Note that all of the primary haul costs are nearly
    Facility costs include the annual transfer station costs ($351,000

equal, and that all but the secondary haul cost for Alternative 1 are nearly equal (the secondary haul for Alternative 1 is much longer and involves more tonnage). Therefore, from a total transport cost perspective, there are economies in the locations of the resource recovery facilities.

The "Facility Costs" and "Revenue" columns of Table 6.1 are also revealing. Alternative 1 has the lowest facility cost by a wide margin, but there are no offsetting revenues. Even with offsetting revenues, the other three resource recovery alternatives have higher costs. From purely a facility cost versus revenues prespective, therefore, the landfill alternative is more economical.

A life-cycle cost analysis appears to be justified because Alternative 2 has an NSC only about 15 percent higher than Alternative 1. Landfill costs will escalate in the future, but the net costs for Alternative 2 may not escalate as quickly because the energy revenues are likely to escalate at a higher rate than the other components of the NSC. There may, therefore, be a cost "cross-over" at some future date between Alternatives 1 and 2. The following LCC analysis was performed to discover if this cross-over might happen and its effect on alternative system ranking when life cycle costs are calculated.

B. Life Cycle Cost Analysis

Life cycle cost calculations were performed for each Alternative system utilizing equations similar to those presented previously in Section III of this chapter. The following factors were used in this analysis:

1. The base-year NSC components shown on Table 6.1 were used as the basic costs.

2. The long-term inflation rates chosen for escalation of costs after 1985 are; 7 percent per year for all transport costs and O&M cost items, 9 percent per year for all steam revenues, and 6.2 percent per year for all electric power revenues.

3. A discount rate of 9 percent per year was used for all resource recovery alternatives and 8 percent per year for the Landfill Alternative (these rates are equal to the bond interest rates).

4. Factors chosen for sensitivity analysis include the O&M inflation rate and the steam revenue inflation rate.

Figure 6.5 displays the results of a present worth (PW) LCC cost calculation for each of the four alternative systems using the factors given previously.

The system with the lowest LCC is Alternative 2--Steam to RCI. Alternative 3--Co-Generation also has a lower LCC than Alternative 1-- Landfill. The conclusion is that under the assumptions made in this LCC analysis, Alternatives 2 and 3 are less costly than landfilling over their operating life span, even though their first year costs are higher than landfilling. Furthermore, the LCC for Alternative 2 is almost half as expensive as for Alternative 1.

The reason for this switch in cost ranking from base year to LCC is that while Alternative 1 has no offsetting revenues, Alternatives 2 and 3 have steam revenues which escalate at 9 percent per year offsetting O&M costs which escalate at only 7 percent per year. The NSC for Alternatives 2 and 3 do not, therefore, rise as quickly as the NSC for Alternative 1.

C. Sensitivity Analysis

If the assumptions made in the LCC analysis are not correct, the results may be different. The following sensitivity analysis was performed to discover if the ranking of Alternatives 1 and 2 would change under different assumptions. In this analysis it was assumed that all transport and O&M costs would escalate due to inflation at 9 percent per year instead of the 7 percent per year used in the LCC analysis. Also, the steam revenue escalation rate was changed from 9

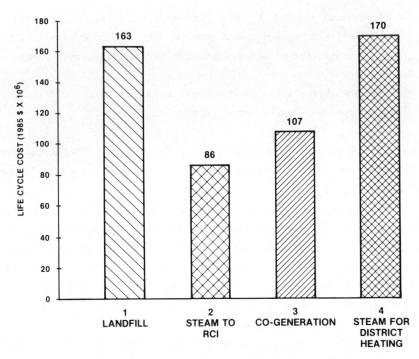

Figure 6.5. River City LCC Analysis, Four Alternatives

percent per year to 7 percent per year. These changes affected the
LCC of both Alternative 1 and 2 as shown in Figure 6.6. Higher O&M
and transport cost escalation resulted in increased costs to both
Alternatives. A lower escalation rate for steam revenues further
increased net costs for Alternative 2. Figure 6.2 shows, however, that
although the life cycle costs are closer in magnitude, the LCC for
Alternative 2 remains lower.

It appears, then, that Alternative 2--Steam to RCI should be
pursued further into a more extensive analysis. Negotiations with RCI
for a steam sales agreement should commence along with a more
detailed cost analysis and the selection of a financing plan for the
required capital (approximatley $100 million). These, and other
implementation planning issues, are discussed in the following chapter.

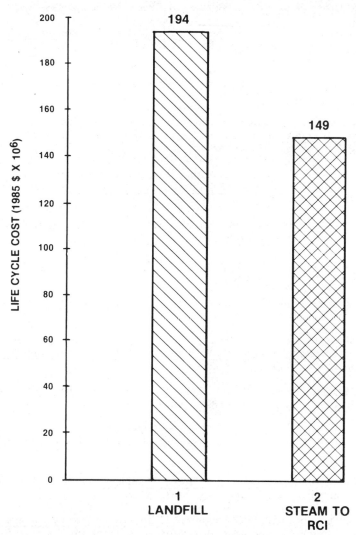

Figure 6.6.   River City Sensitivity Analysis, Alternatives 1 and 2

REFERENCES

1. George F. Haddix, "Regional Solid Waste Problems Two Cases," in Comput. & Urban Soc., Vol. 1, Permagon Press, 1975, pp. 179-193.

2. G.F. Haddix and M.K. Wees, Solid Waste Management Planning Models with Resource Recovery, Proc. ORSA-TIMS Meeting, Las Vegas, 1975.

3. W.J. Baumol and P. Wolfe, "A Warehouse Location Problem," Operations Research Vol. 6, 1958.

4. J.F. Hudson, D.S. Grossman, and D.H. Marks, Analysis Models for Solid Waste Management, Dept. of Civil Engineering, Massachusetts Institute of Technology, 1973.

5. D.H. Marks and J.C. Liebman, "Location Models: Solid Waste Collection Example," Journal of the Urban Planning Division, Proc. ASCE 97(VPI), 1971.

6. K. Spielburg, "Algorithms for the Simple Plant-Location Problem with Side Conditions," Operations Research, Vol. 17, 1969.

7. S.I. Gass, Linear Programming Methods and Applications, 3rd Ed., McGraw-Hill, New York, 1969.

8. I.S. Hillier and G.J. Lieberman, Introduction to Operations Research, 2nd Ed., Holden-Day, New York, 1974.

9. S.H. Russell and M.K. Wees, "Life Cycle Costing for Resource Recovery Facilities," in Proc. 1980 National Waste Processing Conf. ASME, May 11-14, 1980, pp. 259-267.

10. N.E. Mascio, "Predict Costs Reliably Via Regression Analysis," Chemical Engineering, Feb. 12, 1979, pp. 115-121.

11. U.S. Department of Energy, "Proposed Methodology and Procedures for Life Cycle Cost Analysis of Federal Buildings," Federal Register, April 30, 1979.

12. R.S. Brown et al., Economic Analysis Handbook, prepared for Naval Facilities Engineering Command HQ, June, 1975.

13. R.T. Ruegg and J.S. McConnaughey, Life Cycle Costing Manual for the Federal Energy Management Programs, U.S. Dept. of Energy, April, 1979.

14. U.S. Energy Research and Development Administration, Life Cycle Costing Emphasizing Energy Conservation: Guidelines for Investment Analysis, May, 1977.

15. U.S. Office of Management and Budget, Circular A-94.

16. K.J. Arrow, "Discounting and Public Investment Criteria," in Water Research, A.V. Kneese and S.C. Smith eds., Baltimore, 1966.

17. M.J. Bailey and M.C. Jensen, "Risk and the Discount Rate for Public Investments," in Studies in the Theory of Capital Markets, M.C. Jensen ed., Preager, New York, 1972.

18. J.A. Stockfish, Measuring the Opportunity Cost of Government Investment, Research Paper P-490, Institute for Defense Analysis, March, 1969.

19. E.L. Grant and W.G. Ireson, Principles of Engineering Economy, Ronald Press, New York, 1970.

20. R.J. Brown and R.R. Yanuck, Life Cycle Costing, Assoc. of Energy Engineers, Atlanta, 1979.

CHAPTER 7

IMPLEMENTATION PLANNING

I. INTRODUCTION

The activities described in the preceding chapters of this book bring the investigator to the end of the economic feasibility analysis. At this point, the investigator should have enough information to recommend to the appropriate decision-makers further action on one of the alternative systems.

It is not sufficient, however, to merely recommend "further action." Decision-makers (a city council, a county board, members of a regional authority, or others) must be presented with a specific plan for further implementation action before a decision to proceed can be made. The project will stand a much higher chance of being implemented if a detailed, logical plan of action is presented to the decision-makers, and is used to guide implementation activities.

The purpose of this chapter is to present some of the justifications and impediments to resource recovery implementation experienced in actual projects in the U.S. and the general framework of the implementation approach which has been utilized in actual projects. Although the information presented in this chapter can be used by the investigator as a guide, it must be recognized that implementation planning must reflect the specific circumstances and unique forces which exist in each community. General guidelines cannot possibly anticipate all situations which may arise. With the knowledge of the specific community situation and the information presented in this chapter, the investigator should be able to develop a workable, cost-effective implementation plan for presentation to the appropriate decision-makers.

II.   RESOURCE RECOVERY IMPLEMENTATION: JUSTIFICATIONS AND IMPEDIMENTS

Before discussing implementation approaches, it is instructive to consider some of the justifications or motivating factors behind a resource recovery project. Some of the impediments to implementation, or factors which have delayed or stopped projects in the past, are also important to understand when preparing an implementation plan. Much of the following discussion was taken from a paper presented at the National Solid Waste Management Association (NSWMA) conference in May 1981 (1).

A.   Driving Forces

There are a number of means by which resource recovery programs are justified for a particular community (i.e., a city, county or regional entity).   The most common reasons include a) economics (that is, disposal at less cost), b) environmental improvement; and, c) the use of a renewable energy source.   However, while these are reasons which may be used to justify pursuing resource recovery, anything that satisfies the needs of the "driving force" will suffice.   In every successful project there has been some major driving force (a key person, or entity) which causes the project to be initiated.   If the driving force remains in place and has adequate motivation, it can drive the project from initial studies through implementation and construction of a facility.   Many times, however, the driving force changes or loses interest for any one of a number of reasons and as a result, many projects started in the U.S. have either been stopped or have had a "stop-and-go" history.

The major driving forces can be categorized as follows: a) individual ambition/dedication, b) regulatory forces, and c) private sector activity. The following paragraphs are a discussion of each.

1. Individual Ambition/Dedication

Frequently, the impetus for a local resource recovery project originates with the desire of a local individual to pursue resource recovery for personal or professional reasons. Such individuals might include the following:

a. <u>Solid Waste Manager</u> - The person responsible for collecting, and disposing of solid waste in a community (i.e., the local public works director or solid waste superintendent) is frequently the driving force for a resource recovery project. This person views resource recovery as a solution to his problems. Very likely he is faced with landfill depletion, and the prospect of extreme difficulty in siting a new facility. This person may also be looking for a new, less costly method of disposing of refuse. Under the right set of circumstances, resource recovery can be competitive with-- or less costly than--present and projected landfill disposal costs, particularly if haul costs to new, remote disposal sites are considered.

b. <u>Local Planners</u> - Another individual who might initiate a resource recovery program is the local or regional planner. A number of projects have grown out of the planning process. Planning for resource recovery as a long-range alternative is seen by such individuals as a logical opportunity to expand or continue their programs, possibly with the influx of new revenue from state or federal sources. It is unfortunate, but true, that most agencies of this type are not in a position to implement projects, thus a great many of the studies started by this driving force ultimately are not implemented because there is no mechanism for continuing the project beyond the planning stage.

c. Local Politicians - Some projects are the offspring of local politicians who accept the role of the driving force for various personal or professional reasons. Resource recovery is often perceived as a glamorous project, and it is often the largest public works project undertaken in a given area to date. Therefore, if it is successful, there are certain rewards to be reaped by politicians. The project might even be the springboard to new, more ambitious, or higher-level endeavors. There is also a certain amount of peer pressure which local politicans may feel when other communities of comparable size are pursuing resource recovery and the same is not being done in their community.

d. Environmental Groups or Individuals - Local environmentalists or environmental groups have also prompted a number of resource recovery studies. In some, the impetus comes from resource recovery being a current national issue while in others, it is simply a sincere desire on the part of an individual or group to see resource recovery analyzed as an alternative to what they perceive as environmental degredation from the landfilling of raw refuse.

Whatever the motive, and no matter how lofty the ideals, the unfortunate fact is that such groups are not in a position to implement projects. Thus, although they may cause resource recovery to be studied, implementation is not assured.

2. Regulatory Forces

A large number of projects around the country have resulted from either state or federal action. Some states have passed laws requiring counties or regional/metropolitan agencies to develop solid waste management and resource recovery programs. This has forced many communities to study resource recovery as a solid waste management alternative. In addition, recent federal legislation has had the effect of making resource recovery more cost competitive with continued land disposal of refuse. Pertinent legislation includes the Resource Convservation and Recovery Act (RCRA), the Federal Energy Act (specifically those sections relating to the Public Utility Regulatory Policies Act--PURPA, and oil entitlements), deregulation of natural gas and oil, and other new solid waste resource recovery activity by the U.S. Department of Commerce.

One of the primary products of RCRA is the local requirement to either close or upgrade existing land disposal facilities which do not meet stringent criteria for design, construction, and operation. The result is an increase in the cost of disposing of refuse by landfill, thus partially closing the gap on the cost of landfill relative to the cost of resource recovery programs. The PURPA provisions of the Federal Energy Act require utilities to purchase power generated from refuse and other biomass facilities at a price which enhances the economic viability of resource recovery projects. When combined with the higher cost of natural gas and fuel oil which has resulted from deregulation of these fuels, the net effect of all of these federal actions is to make resource recovery considerably more attractive from an economic perspective. While the legislation itself is not a major driving force, it affords other local forces an opportunity to assume this role with an improved prospect for ultimate success.

3. Private Sector Activity

There are a number of forces within the private sector which may initiate resource recovery activity at the local level. Among these are vendors of solid waste resource recovery systems, who present the convincing argument that if the local community will simply guarantee the waste stream to the vendor at a reasonable tipping fee, the vendor will relieve the community of the problems which have plagued them in getting rid of solid waste. Similarly, this category also includes consulting engineers who are also always looking for new or continuing opportunities for work. Another segment of the private sector that has been responsible for certain projects are the high technology firms from the aerospace and electronic industries which have spun off into the resource recovery field and have convinced local decision-makers to allow their community to be used for product development. Finally, there is the recent movement into this relatively new field by the banking community, which is always looking for new investment opportunities.

While all of these private sector companies and/or personalities may be considered driving forces for making resource recovery happen in a specific local circumstance, it must be recognized that they cannot, by themselves, cause it to happen. They must utilize one of the previous major driving forces as an ally or vehicle.

B. Impediments

Because resource recovery involves the cooperation of many different elements of a community, there are many impediments to implementation. These impediments can be classified into a number of different categories as given in the following paragraphs.

1. Lack of Local Commitment

Projects which do not have a major commitment of local funds many times stop when the state or federal monies which have funded the project run out. Conversely, those projects which have a significant commitment of local funds frequently proceed in spite of major funding impediments. In addition to a major commitment of local funds, there must also be a strong personality or lead agency determined to make the project go and which will allocate the necessary personnel and other resources to cause it to happen. Lacking either of the these requisites, the project is very likely in serious trouble, even though it may not be readily apparent in the early stages.

2. Lack of Key Elements

There are five key elements which must exist for a successful resource recovery project:

1. A suitable site for the facility
2. Problems with current landfill disposal, i.e., high cost, or space depletion

3. An available, willing energy buyer
4. Control of the waste stream, or a commitment to
    deliver waste to the facility
5. A local lead agency which has the ability and desire to
    implement the project

The lack of any one of these key elements can cause a project to fail.

3. Unfavorable Economics

The major deciding factor in implementation of any major public works project for which there are competing alternatives, or which must compete with other local projects for funding, is economics. Unless the resource recovery project can demonstrate economic viability relative to other waste disposal alternatives, it will likely not be constructed--nor should it be.

4. Institutional Issues

This category is perhaps more responsible for determining a project's fate than any other. It is undoubtedly the most difficult category to resolve, and is also the area that causes the most problems for consultants, vendors and others who are trying to work with the community, because it is an area in which they have little or no control.

Because resource recovery projects are by their very nature large, capital-intensive endeavors, they can be perceived as either major boondoggles or as a large "jar of jelly beans" of which all entities want to get their fair share. In all too many cases, projects suffer because the participating entities get caught up in jurisdictional battles which are completely unrelated to solid waste or resource recovery issues. For example, in one actual project, the trade-off for allowing one entity to become the lead agency was that the other entity was allowed to operate the city/county building. Other examples of trade-offs are new fire stations, expansions of existing water or wastewater systems, land purchases, payments in lieu of taxes, reduced rates for waste disposal or for energy purchase, and a myriad of other items. All of these can be attributed to the human tendency to "look out for Number One.". Unfortunately in projects of this magnitude, these jurisdictional battles exert a major economic penalty on the project. It is not unusual to suffer a project construction cost escalation of $25,000 to $50,000 per day while the institutional issues are being resolved.

5. Prolonged Project Development

This impediment is closely related to Item 4. At the outset of most resource recovery projects, there is a great deal of enthusiasm in all parties associated with the project. Unless the project is expedited, it becomes extremely difficult to maintain this level of interest and enthusiasm, especially when projects extend over several years. The time required for project development thus far in the United States is typically five to eight years for those projects which have moved along without any major problems. Other projects have been developing for 10 to 12 years. During this time, costs go up, elections occur and players change, all of which can lead to diminished interest in the resource recovery program in favor of some new project which captures the community's interest.

6. Legal Barriers

There are a number of legal barriers which can impede the development of a resource recovery program. Control of the waste stream is one critical issue which can, and has, stopped projects. Other legal barriers include limitations on contracting/financing which impede some lead agencies. Specifically, it is necessary that the lead agency have the authority to enter into long-term contracts for waste supply, long-term contracts for facility operations, and long-term contracts for sale of energy and other products. It is also necessary that they be able to enter into such contracts without being required to periodically open them to competitive bidding. In many projects, these types of limitations have required the local community to seek new legislation to enable them to eliminate these legal barriers. This seeking of new legislation is also an impediment because of the time it takes to develop new legislation and get it passed by the state legislature or assembly.

7. Social/Environmental Issues

While meeting essential social and environmental criteria in the development of resource recovery programs is necessary, it must be recognized that these issues are an impediment to timing and may place unreasonable demands on the project. Some of the specific issues which can be especially time consuming are facility siting, air quality monitoring, and the permitting/hearing process. It is not uncommon for the permitting process to take a year or more. It is also not uncommon for siting issues to also take over a year, and in some cases to stop the project all together. Other lesser environmental issues which can usually be resolved in a much shorter time frame are increased traffic, potential odors and perceived noise problems.

8. Industry Failures

A major impediment which all current and future projects must overcome are the number of resource recovery facilities in the U.S. which have either failed from a technical standpoint or are in serious economic trouble for technical or other reasons. Each new project bears the burden of proving that it is not similar to these failures. Because the failure rate has been quite high to date, there is a general, well-founded "fear of the unknown" among local decision-makers; however, it is encouraging to note that there are facilities which run quite well at a cost which is competitive with other alternatives. These facilities employ proven technology and are in localities in which recovered energy value is high and the costs of competing landfills are also high.

9. Competition

An impediment which has emerged only recently is the possibility of
competing facilities (or projects) initiated by regional or state
authorities, by utilities (local, state, or regional) or by other private
developers. To date, no project is known to have been stopped by this
manner of competition, but several have been delayed. While
competition may be an impediment to the project developer, it can also
be viewed as a hopeful sign. The emergence of competition signifies
that resource recovery may finally be reaching the status of individual
project economic viability. If the competition continues to develop, it
may complicate the institutional issues which must be resolved, but it
will also widen the spectrum of project sponsors and types of recovered
energy buyers. For these reasons, competition may be an
implementation impediment at the local level, but it may act to
increase the number of resource recovery projects nationwide.

III. IMPLEMENTATION APPROACH

There have been several guides written for implementation planning.
Some are relatively straightforward (2), while others are highly complex,
attempting to account for all of the possible situations which may arise
in a given community (3). While these reference materials may be
valuable in showing implementation approaches, the investigator should
be wary of resource recovery implementation "cookbooks." Some
persons responsible for implementation planning in certain communities
have made the mistake of trying to fit a local situation into the
highly-structured implementation framework given in certain references,
instead of the reverse. As a result, many projects have become
bogged-down in unnecessary tasks which act to delay implementation.

Although an implementation plan must be developed to fit the
particular situation in a given community, there are certain features
which should be part of any implementation plan. It is the purpose of
this section to briefly describe these features. Actual project

experience and the implementation planning reference materials indicate that a community cannot move directly from a feasibility study into procurement of the selected system alternative. An intermediate planning step is needed. This step or phase has been called the "key decisions" step, "planning" phase, "procurement planning" phase, and "system configuration" phase. For the purpose of this discussion, the term "system configuration phase" will be used. Project experience and the implementation reference materials also indicate that there is a separate "procurement" phase which follows this intermediate step. Steps beyond procurement include actual construction, start-up, and full-scale operation. The following sections describe some of the features which should be incorporated in these implementation phases.

A. System Configuration Phase

If the feasibility study has been performed following the guidelines given in this book, the output is a recommendation of one system alternative. The information provided in the feasibility study about the recommended system alternative includes a generalized site location, an estimate of how much waste is to be handled by the system, a selected resource recovery technology with an initial estimate of capital and annual costs (based on an initial assumption of the financing method), and the selection of the most desirable energy buyer with whom preliminary discussions have been held. Note that much of this information is preliminary in nature (as it was intended to be for the feasibility study) and cannot, therefore, form the basis for immediate system procurement. Certain incomplete and/or improperly prepared feasibility studies have provided even less information. For instance, some feasibility studies have ended without settling the question of which technology to utilize, which energy market is most desirable, or where the supporting landfill is to be located.

The purpose of the system configuration phase is to provide an intermediate step between the feasibility study and the procurement activity. The following tasks should be accomplished in the system configuration phase: 1) select the lead implementation agency, 2) establish control of the waste stream, 3) negotiate a preliminary agreement with the energy market, 4) select a specific site, 5) establish a financing and procurement plan, 6) perform a preliminary environmental analysis and permits investigation, and 7) confirm economic feasibility.   There is disagreement on how these tasks should be sequenced and how much time is required to accomplish this phase. Some reference materials (3) indicate a task sequence which requires up to 19 months to complete.   In actual practice, however, these tasks have been accomplished in certain circumstances within 90 to 120 days (4).   The impetus for expiditing the process as much as possible is the rapid rate of construction cost escalation.   For example, with a $100 million project (a moderately sized project), escalation at a rate of 12 percent per year adds an average of about $33,000 per calendar day of delay to the capital cost of the project.

The investigator should choose the task sequencing scheme and timing appropriate for the local situation.   It is important to show these details to local decision-makers in the implementation plan.   A written narrative of the required activities is one way to make the presentation; however, a combination of a written narrative and a logic diagram is usually more illustrative.   The example implementation plan for River City in Section IV of this chapter utilizes one kind of logic diagram which can be useful.

The following paragraphs describe in more detail the tasks which should be part of this implementation phase.

1. Select Lead Implementation Agency

The primary purpose of this task is to select and fund an organization with the powers, resources and desire to implement the recommended project.   This task may be relatively simple if an existing agency such as a city or county government has agreed to be the lead agency. However, the formation of a new implementation agency, such as a regional authority or a non-profit public corporation, will be much more complicated.   In addition, jurisdictional battles may arise over control of the project which can further complicate this task. The project cannot move into the procurement phase before an effective implementing organization has been formed or designated and funded. Indeed, effective negotiation with energy markets cannot proceed without this selection. The object is to select an implementation agency which can obtain control over waste disposition in the study area, has the power to obtain capital financing for the project and enter into contracts for its construction and/or operation, has the power to enter into contracts for the sale of recovered energy and materials, and has the political power to obtain the required approvals from all project participants (in multijurisdictional approaches). The options include: city, county, or state governments, a solid waste authority, a non-profit corporation, mulitcommunity cooperatives, a special district. The reader should consult reference 5 for details of the advantages and disadvantages of each option.

The plan should show all of the steps necessary to establish the agency, including: obtaining resolutions of the participating municipalities, obtaining passage of any required local or state legislation, establishing the range of powers and services to be given to the agency, investigation of agency funding options, and any other required steps.

Experience has shown that successful implementation requires a full-time project director and a supporting team of technical, legal, financial, and public relations experts (6,7). A project team will have been established to coordinate the feasiblity study; hence, it is logical to utilize this existing team as much as possible. If a full-time director has not already been established for the feasibility study, such a director should be hired either as a municipal employee of one of the municipalities participating in the project (if more than one is involved) or by a regional agency responsible for this phase of implementation. Someone of high professional status, who will report to the highest levels of government, is needed for this position. The supporting team can be a combination of government employees and consultants.

The plan may also call for the formation of a citizens' committee (if not already established) composed of representatives of environmental groups, local industry and labor organizations, local chambers of commerce, private waste haulers, neighborhood groups in the areas of potential sites, and potential energy customers. Such committees have been helpful in the implementation of certain projects if given a decision-making role.

2. Establish Waste Supply Control

The control of the waste disposition in the study area is critical to the success of a resource recovery project; however, it may not be necessary to control all of the waste generated in the study area if the facility needs only a portion. The implementing agency can obtain such control by several methods, including: obtaining delivery agreements from municipalities in the study area which have municipal collection service or which contract for collection, obtaining agreements from all municipalities in the study area to close municipally-owned landfills to private haulers, establishing a drop fee incentive to private haulers for depositing waste at the desired location, passing local or state legislation requiring all private haulers as a condition of their license to deposit waste at only certain locations. Some financing methods will require strong waste supply control measures (such as legislation) in order to successfully raise funds for construction. As a result of this task, an update and confirmation of design waste capacity for the facilities in the selected system can be made.

3. Negotiate Preliminary Energy Purchase Agreements

Another major activity in this phase is to obtain a preliminary energy purchase agreement from a recovered energy buyer. Projects which proceed without such an agreement have generally not been successful in the past. It is not necessary to negotiate a final contract with the energy buyer at this point, but a firm letter of intent which contains the preliminary conditions of purchase in sufficient detail for planning purposes should be obtained before the project moves further. If it is not possible to obtain a preliminary agreement with the energy buyer selected in the feasibility analysis, other buyers should be contacted.

This change in energy buyer may require changes in the site and/or technology of the resource recovery facility which, in turn, will change the net system cost.

4. Select Site

The plan should include this step as an essential task. Certain resource recovery projects have been delayed solely because of disagreement over the proposed site. The plan should show the use of input from the citizens' committee in the selection of a specific site. Some projects have utilized a rating and ranking scheme with members of the citizens' committee to locate the best site (8,9). By utilizing the citizens' committee in this matter, future disagreement over the siting question may be avoided, but it should be recognized that the site selection process will consume more time with such an approach.

5. Preliminary Environmental Analysis

A preliminary environmental analysis and permits investigation should also be part of the plan for this phase. This task should include an examination of the potential environmental impacts associated with the construction and operation of the proposed facilities on the selected site or sites, along with an identification of the major environmental permits and associated studies required for construction and operation of the facilities. By performing such an analysis at this point in the implementation, project planning can be done to avoid construction and start-up delays, and to minimize environmental impacts.

6. Select Financing and Procurement Plan

The plan should call for the examination of the range of financing and procurement options and the selection of a procurement phase plan. The chosen plan will depend on the desires of local decision-makers and the financial strength of the implementing agency. Options for procurement include variations of three major types:

architect/engineer, turnkey, and full service. The reader should consult reference 2 for more detail about these methods. The procurement method chosen in this step will determine the steps necessary in the procurement phase.

Experience has shown that the complexities of resource recovery financing requires the expertise of specialists in the financing of such projects. The plan should, therefore, call for retaining a financial consultant to examine the possible financing mechanisms available to the implementing agency for the project under consideration, and recommend the best method. Note that if the financing method selected is significantly different from the method assumed in the feasibility study, then the economic analysis must be revised.

7. Confirm Economic Feasibility

The plan should call for this step after the preceding tasks have been accomplished. The economic feasibility calculations may require revision due to changes in waste quantities resulting from waste supply agreements, changes in energy market value as a result of preliminary energy purchase agreements, a different site as a result of the site selection activity, or a change in financing method. A change in the availability of landfill space may also affect economic feasibility.

B. Procurement Phase

Following a positive decision by the implementing agency after confirmation of economic feasibility, the object of this phase is to finalize arrangements for procurement, secure the required permits; prepare a final cost estimate, finalize waste supply and energy purchase agreements, finalize the financing, and to design, construct and start up the required facilities. The specific activities in this phase cannot be known until the results of the financing and procurement planning activities in the system configuration phase are known. However, the investigator can present a plan which contains enough information to give decision-makers a basic knowledge of the required activities.

The time required to move from the beginning of this phase to the start of construction may take as long as 39 months according to certain references (3), but has been accomplished in as few as twelve months (4). The time required depends on the procurement method chosen and the length of time necessary to obtain permits in addition to other factors. The salaries and office expenses of the implementing agency plus consulting fees (which can be significant in this phase), can be covered out of financing proceeds. All design, construction, and start-up activities will also be paid by financing proceeds after the final financing. Some of the major activities which are part of this phase are described in the following paragraphs.

1.   Pre-Design Activity

The required facilities must go through a conceptual design review process so that construction and environmental permits can be obtained, and so that a final cost estimate can be prepared. Depending on the procurement method chosen, the activities in this step may involve the selection of an architect/engineer who will prepare the preliminary design, or the selection of a turnkey or full-service contractor who will prepare the preliminary design. Instead of showing one procurement method in the plan, the investigator may wish to display all options.

2.   Secure Permits

For a resource recovery facility, the most critical and time-consuming permit to obtain is the air quality permit. If air quality monitoring data are required to establish background pollutant levels, one year of ambient air quality monitoring may be required. Water quality and solid waste permits will also be required. The implementation plan should show the activities required to obtain these permits as a critical step in this phase.

3. Final Cost Estimate

A final capital cost estimate from the contractor or the architect/engineer must be made before financing. This cost estimate based on the preliminary design must be precise because the capital financing will be based on this final estimate. An independent cost estimating consultant is sometimes utilized to confirm the final estimate.

4. Finalize Waste Supply and Energy Purchase Agreements

Before financing can commence, binding agreements for waste supply and energy purchase must be obtained. Detailed terms and conditions must be negotiated with the energy buyer. If preliminary agreements have been well-defined, this step should only be a matter of working out details. However, problems with contract terms can cause delays in this phase. Indeed, some projects have been stopped altogether because of the inability to obtain an acceptable final energy purchase agreement even after a preliminary agreement has been made.

5. Final Decision Point

The plan should show a final opportunity to terminate or delay the project before authorizing the financing. Reasons to terminate or delay may include the inability to secure required permits, a final cost estimate which is too high, or the inability to negotiate a binding contract for the sale of recovered energy.

6. Financing

The plan should call for a financial "closing" (bond sale or other method) following a positive decision at the final decision point. If the sale of bonds is the financing method selected, an underwriter, bond counsel, and (in certain financings) an engineer must be retained to draw up the prospectus and other documents necessary to market the bonds.

7. Design, Construction, and Start-up

These activities should be shown in the plan following the financing
step. It is important to show a start-up step because this activity can
require a significant amount of time and money for a resource recovery
facility.

IV. RIVER CITY IMPLEMENTATION PLAN

The analysis for the River City example in Chapter 6 indicates that
Alterantive 2--Steam to RCI should be further pursued. This
alternative involves two transfer stations and a mass burn resource
recovery facility which produces steam for sale to an industrial
customer RCI. The implementation plan for this project, presented in
the following paragraphs, is designed for presentation to the city
council for approval. Note that the plan shows both an approximate
dollar and time commitment at each phase.

A. System Configuration Phase

Figure 7.1 represents the tasks involved in this phase of
implementation. The tasks begin with the selection of a lead
implementation agency and a project team. In the River City case, it
is assumed that the city wishes to be the lead agency, therefore these
steps are relatively simple. The plan then moves to a number of tasks
undertaken simultaneously. The plan shows that the city (perhaps with
the assistance of consultants) would: establish waste supply control (by
a method to be investigated), negotiate a preliminary steam sales
agreement with RCI, select the sites, draw preliminary site layouts (for
the transfer stations and resource recovery facility), develop a financing
and procurement plan, and perform a preliminary environmental analysis.
Economic feasibility is then re-analyzed with updated information. The

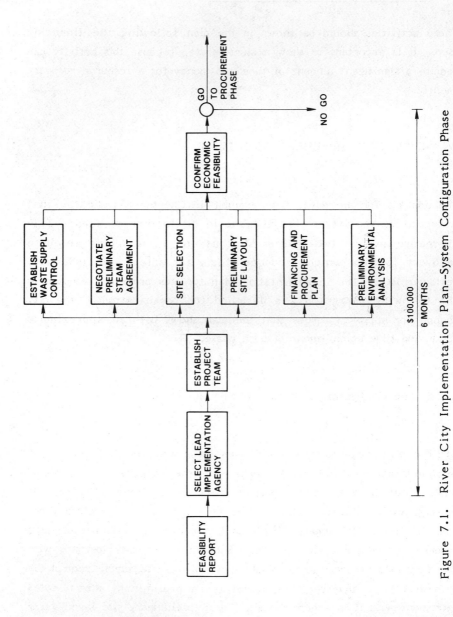

Figure 7.1.   River City Implementation Plan--System Configuration Phase

plan shows a decision point to be made by the City Council for continuation into the procurement phase. The plan calls for an initial project funding for this phase totaling $100,000, which covers project team operating expenses and consulting fees over a six month period.

B. Procurement Phase

Figure 7.2 illustrates the tasks in this implementation phase. Because the procurement method to be chosen is yet unknown, the plan shows both the architect/engineer (A/E) approach and the full-service approach as possibilities. The securing of environmental permits, a final cost estimate, and the negotiation of a final steam contract with RCI are also shown as primary activities. The city council must make a final decision before the financial closing date (a bond sale is indicated in this plan). Final design, construction, start-up, and full-scale operation are also shown in the plan. Depending on the procurement method chosen, a range of $500,000 to $2 million is shown for project team and consultant expenses following the system configuration phase. The full-service approach will cost the city less initially, but the selected contractor will expect to recoup pre-design costs after selection. With the A/E approach, all pre-design costs are incurred before the bond issue. The plan also indicates that $100,000 and six months will be required to market the bonds (a revenue bond issue is assumed) to pay for legal, financial, and engineering analyses of the project. During the construction period, expenses for project administration and construction costs will be paid out of bond proceeds.

With a 36-month construction period, the total elapsed time shown between the decision to proceed past the feasibility study and the start of full-scale operation is 66 months (5 1/2 years) with a dollar outlay (exclusive of construction costs) of between $700,000 and $2.2 million. Such a commitment is not unusual for projects of this nature. It is important that the decision-makers are aware of the size of this commitment prior to commencing implementation.

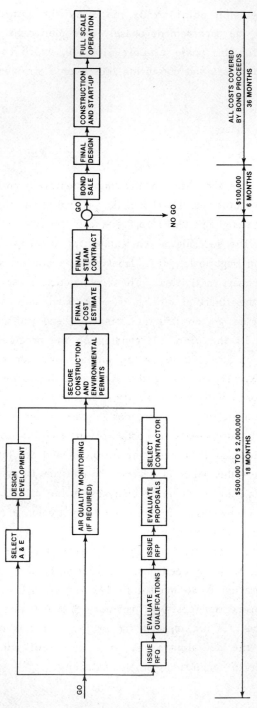

Figure 7.2.   River City Implementation Plan--Procurement Phase

REFERENCES

1. F. A. Borchardt, Resource Recovery--Justification/Impediments, National Solid Waste Management Association Conference, panel discussion, Dallas, Texas, May, 1981.

2. Mitre Corp., Resource Recovery Plant Implementation, Guides for Municipal Officials: Procurement, U.S. EPA Publication SW-157.5, U.S. Government Printing Office, Washington, D.C., 1976.

3. U.S. Environmental Protection Agency, Resource Recovery Management Model, Pre-publication Copy, September, 1979.

4. Henningson, Durham & Richardson, Inc., Hillsborough County Resource Recovery Project, System Configuration Report, May, 1981.

5. K. Anderson, et al., Decision-Makers Guide in Solid Waste Mangement, 2nd ed., U.S. EPA Publication SW-500, U.S. Government Printing Office, Washington, D.C., 1976.

6. C.A. Johnson, Resource Recovery Decision-Makers Guide, National Solid Waste Management Association, Washington, D.C., 1979.

7. A. Shilepsky and R.A. Lowe, Resource Recovery Plant Implementation: Guides for Municipal Officials - Planning and Overview, U.S. EPA Publication SW-157.1, U.S. Government Printing Office, Washington, D.C., 1976.

8. Henningson, Durham & Richardson, Inc., City of Charlotte, North Carolina, Solid Waste Disposal and Resource Recovery Study--Vol. II, Potential Facility Locations, 1978.

9. Henningson, Durham & Richardson, Inc., Southern California Urban Resource Recovery Project: Vol. II--Resource Recovery Guidelines and State-of-the-Art Assessment, February, 1976.

GLOSSARY

ACRS - Accelerated Cost Recovery System

Btu - British Thermal Unit

C - Centigrade

CY - Cubic Yard

d-RDF - Densified Refuse Derived Fuel

EPA - United States Environmental Protection Agency

ETC - Energy Tax Credit

F - Farenheit

F.O.B. - Free on Board. If a price is stated F.O.B. seller
location, buyer pays for and accepts responsibility for shipment. If
a price is stated F.O.B. buyer destination, seller pays for and
accepts responsibility for shipment.

HMS - Heavy Media Separation

IDC - Interest During Construction. This term refers to the
interest to be paid by the owner on bonds or loans during the
construction period of a facility.

IRS - United States Internal Revenue Service

ITC - Investment Tax Credit

KW - Kilowatt (1,000 watts). Rate of electric power generation.

KWH - Kilwatt-hour. Quantity of electric power equal to that
expended by one kilowatt in one hour.

lb - Pound

LCC - Life Cycle Cost

LFl, LF2 - Locations of potential landfill sites in the hypothetical community, River City.

Mlb - One Thousand pounds

MW - Megawatt, One Million watts

NCRR - National Center for Resource Recovery

NSC - Net System Cost

O&M - Operation and Maintenance

pph - Pounds per hour

psig - Pounds per square inch gage (pressure above atmospheric pressure).

PURPA - Public Utilities Regulatory Policy Act

PW - Present Worth method of life cycle cost analysis.

RCI - River City Industries, a hypothetical industry in River City.

RCPL - River City Power and Light, a hypothetical electric power utility in River City.

RCRA - Resource Conservation and Recovery Act of 1976, PL 94-580.

RDF - Refuse Derived Fuel

RRl, RR2 - Locations of potential resource recovery facility sites in the hypothetical community, River City.

scf - Standard cubic foot (at 68 degrees F, and a pressure of 29.92 inches of mercury).

SIC - Standard Industrial Classification, U.S. Office of Management and Budget

T1, T2, T3, T4, T5 - Locations of potential transfer station sites in the hypothetical community, River City.

ton - Two thousand pounds

ton - A unit of cooling capacity equivalent to 12,000 Btu per hour.

TPD - Tons per day

TPW - Tons per week

UAC - Uniform Annual Cost method of life cycle cost analysis.

WGD - Waste Generation District

INDEX